The Road Towards *DemocraCity*

Marco Carcea

The Road Towards *DemocraCity*

A Creative Approach to Transition in a Post-Development Based Society

palgrave
macmillan

Marco Carcea
Comune di Crotone (City Hall of Crotone)
Crotone, Italy

ISBN 978-981-16-3504-5 ISBN 978-981-16-3505-2 (eBook)
https://doi.org/10.1007/978-981-16-3505-2

This Palgrave Macmillan imprint is published by the registered company Springer Nature Singapore Pte Ltd.
The registered company address is: 152 Beach Road, #21-01/04 Gateway East, Singapore 189721, Singapore

to Mamma and Pà. Thank you for believing in me...
and to C. ... None of the wonders that I have experienced
in the last 10 years would have happened without you ......

Contents

CHAPTER 1

Introduction

For the last ten years, the world economic system has been suffering from the effects of the Global Financial Crisis (GFC) and the ineffectual regulation of the world's financial systems. A second significant crisis occurring at the same time is the environmental crisis due to the increasing threat posed by climate change. These events have drawn more attention to the issues of sustainability and the impossibility of infinite economic growth in a finite world. A significant paradigm to emerge out of these crises is the idea known as "Degrowth Theory" introduced by Serge Latouche in 2007.

This paradigm advocates the need for a transition to a more "sustainable" social, economic, and environmental system. Within the framework of Degrowth Theory and combining it with the application of the concept of resilience for communities (Kimhi & Shamai, 2004), this thesis investigates the existing literature on degrowth and resilience within urban centres. The aim of this book is to find possible and achievable approaches discussed in the literature in the broad area of degrowth and resilience that still require comparison and oversight to provide a better understanding of this theoretical framework.

This study combines these two approaches to develop a new perspective and to lay the foundation for the difficult but necessary path for the creation of an inclusive democratic society and to safeguard the planet threatened by the dematerialization of consciousness brought about

M. Carcea, *The Road Towards DemocraCity*,
https://doi.org/10.1007/978-981-16-3505-2_1

by the current economic system. One such multidisciplinary approach, synthesized by Latouche (2007) in the "8 Rs" and combined with the theoretical framework of the "3Rs" of Resilience (Norris et al., 2008), could create a circle of virtuosity, in stark contrast to the already mentioned carousel, characterized by the reintroduction of solidarity and trust. This approach could supplement a new horizontal reciprocity that could replace the hierarchies of the consumer society.

The transition to this new way of thinking represents a major change in the collective imagination and it is, thus, a radical political challenge that offers a new narrative that I have identified in the rise of the *homo resaliens*: the one who has understood the sense of one's limits, the one who is able to arise from an overturned boat (metaphor for our capitalist society). In order to prove the possibility of achieving such an inversion of our collective imagination, Transition Darebin" —part of the global Transition Network—(Melbourne, Victoria), and Fiorivano le Viole (Perugia, Italy) two community projects will be utilized as examples of this experience, examples of the possibility of achieving what we call "*DemocraCity*".

The concepts of economic degrowth, egrowth, resilience, alternative currencies, and Transition Town movements have only recently started to gain prominence in economic, social and policy debates as a response to environmental and social sustainability of economic growth. Degrowth theory represents, from many perspectives, a natural evolution—or the heritage—of the non-superficial trust between humankind and nature and, as emphasized by Bogdanov (1923, cited in Gare, 2000), an expression of human resilience.

Degrowth theory, for the very first time, has focused attention on the issue of compatibility between the functioning of a "civilized world" and the natural biological space. The following sentence used and "abused" by Degrowth scholars such as Latouche (2006), Chochet (2009), Pallante (2007), Jackson (2011) and Kallis (2011), is a recurring mantra in the context of Degrowth. In effect it is not possible to have infinite growth in a finite world. This phrase emphasizes the fact that unlimited economic growth is not sustainable for the Earth's ecosystem.

Degrowth theory sees an urgent need to restore the balance between policy makers, mankind and nature. Degrowth aims to move forward and stresses the pressing need for a radical reversal in the economic and societal directions espoused by the dominant ideology of today, i.e. capitalism. Degrowth, suggests an alternative perspective, one that is in contrast to the several existing development models, one that shifts the target from quantitative growth to qualitative development.

Despite decades of pertinent studies conducted by scholars, biologists, climatologists, such as Godbout (1998), Gare (2000), Beckerman (1974), Belpomme (2004) Besset (2006), Latouche (2007), Barrett (2012), Bornestein (2013), Bloemmen et al. (2015), Muraca and Schmelzer (2017), providing evidence of the impossibility of infinite growth on a finite planet, not one of the economic, social and health objectives set by the international community for 2015 has been achieved.

Despite national and international observers, intergovernmental organizations and other international bodies, confirming that our production and our consumption is overburdening the regeneration capabilities of the biosphere, threatening the health and safety of all living organisms on the planet, there is neither adequate investigation nor critical evaluation of practices employed in the current model of techno-economic, capitalist and mercantilist development. In this model the only acknowledgement of and concession to the looming environmental crisis is the addictive addition of the adjective "sustainable"—possibly a term that has lost its meaning.

Sustainability and development are, according to Latouche (2006), an oxymoron. Sustainability, Latouche argues, attributes a qualitative value to development although development is linked to a concept of quantitative economy and is based on the one-way consumption of resources such as natural ones that are running out. According to Latouche (2007) more "development" as well as "a new" growth can be considered an instance of either great naivety or great hypocrisy. Based on his view, "development" is a toxic, infected word, which has various meanings. A growth society, based on capitalism, technology, and globalization, is based on the presupposition of production-consumption, on what is defined as the "diabolic carousel".

This book addresses the theme of Degrowth, a topic that has been at the centre of numerous varied debates over the last ten years; a debate in which contrasting interpretations, often inconsistent and affected by ideological emphasis, have yet to provide any concrete outcome. Too often the word utopia is present next to the term or definition of Degrowth. It is important to emphasize that Degrowth theory, as a result of economic theories developed in the 1970s, was one of the first social science theories that emphasized issues related to the well-being of the individual, to the interdependencies between the socio-economic system and the natural environment, and to environmental degradation and the finite nature of natural resources.

Degrowth theory is based on the contributions of the theories mentioned in the field of economic growth which are ecologically and socially unsustainable. The theoretical and scientific framework of Degrowth was developed almost 50 years ago. The Degrowth paradigm should be framed as the result of a scientific process that has taken up issues and ancient approaches, which for a long time had been considered marginal and heterodox as discussed in the *The Gift* of Marcel Mauss (1925). This theory, according to Kerschner (2010), Trainer (2012), and Alexander (2012), Andreoni and Galmarini (2014), includes the ideas of voluntary simplicity, reciprocity, conviviality, and primitivism.

However, the provocation of Latouche's Degrowth transcends the concept of utopia. For this reason, the aim of this book is seeking to provide a multidisciplinary and multidimensional conceptual framework. This monograph aims to demonstrate that Degrowth is a feasible and desirable way to rebuild society through micro-changes and rediscoveries.

Terminology

It is important to identify some key points, in our current system, in an attempt to define the meanings of Degrowth and the scope that drives this book:

a. The undefined pursuit of growth is incompatible with a finite planet.
b. A reversal of capitalist society's approach to life, resulting in the abandonment of the objective of unlimited growth whose engine is solely the search for profit by the holders of capital is required.
c. It has been more than ten years since the publication of the *Petit traité of* Latouche and the publication of the *Handbook* by Hopkins, key texts used in this study for the evaluation of the theories under discussion in the search for what has been defined as "*Democracity*".
d. Without a change in the current direction, guided by the idea that economic growth is an end in itself, contemporary society will face disastrous and irreversible consequences for the environment and humanity.

Starting from this last point, it seems necessary to consider that Degrowth theory has its foundation in the criticism of the "gross domestic product" (GDP) framed as an imperfect measure of well-being and from the common view that well-being is quantifiable, measured in units of consumption and acquisition of goods. GDP does not consider the depletion of natural resources, global warming, waste, and the injustices of globalization. GDP is a purely mercantile flow, which not only considers as a positive any kind of production irrespective of its nature but also does not take into account actual contribution to the real individual and collective well-being. We should be aware that any excess growth in gross domestic product corresponds to a decline in life expectancy on Earth.

How many resources and how much work will be required in the future to repair the damage of today? How much food and how much natural wealth are we mortgaging? How much would the economy be growing if we subtracted each year from the GDP the costs of repairing future disasters? Without a doubt, it seems reasonable to argue that GDP would be much lower if countries included the social costs of the damage caused by excessive production and consumption in their calculations, and if natural energy and raw materials consumed today were considered inevitably lost for future generations.

According to Bornstein (2013), reducing the ecological footprint and exploiting natural resources to levels that are compatible with the established capacity of the planet's limits would constitute the ultimate achievement (*eudaimonia*). Only when mankind understands the difference between *well-being* and *well-have,* it can be part of the virtuous circle. Degrowth should be the tool that allows us to escape the mercantile system, replacing GDP with GNH (Gross National Happiness). Such an outcome would be a social revolution, an extraordinary example of resilience that will phase out *homo oeconomicus* (capitalistic man) and it will guide us towards the rise of *homo resaliens* (resilient man).

Economic Degrowth is embodied in an overall reduction in produced and consumed goods, and resources used. This reduction will be achieved through a total transformation of the socio-economic, political, and collective imagination towards a sustainable structure, with the perspective of a significant increase in social welfare.

As will be set out in the analysis of the relevant literature of the theory under discussion, one of the main criticisms of Degrowth theory is that such theory seeks to abolish the very cornerstones of the current dominant ideology based on the idea of unlimited economic growth. This point of view could be considered plausible. The proposal of Degrowth is changing the foundations of the predominant ideologies.

Many scholars of Degrowth theory (Daly, 2010; Jackson, 2009; Kallis, 2011; Kallis et al., 2009, 2010; Kerschner, 2010; Muraca & Schmelzer, 2017) argue that growth proponents lack a realistic perspective and the capacity for reflection; and that they confuse a constructive and articulated criticism of technology, calculating thought and industrialism with a return to pre-modern stage of human history. Degrowth is a series of small moves in the areas of changes in lifestyles, patterns of consumption, and production patterns.

The Road Towards *DemocraCity*

As mentioned above, Degrowth theory is a revolutionary approach which posits that the role of technology should not be underestimated. Technology should not be understood as having a merely technical value, but as a project with the aim of increasing the level of knowledge and reorganizing social complexity, as well as rediscovering the equally complex relationship between mankind and the environment. Society must rethink technology as the last bastion of defence of social and natural capital. Technological progress must not be achieved at the expense of the natural environment and its limits.

Contemporary society must think smart and consider alternative technologies, finding ways to replace Artificial Intelligence (AI) with EI (Environmental Intelligence). There are countless scientific studies that affirm the need to respect the cycles and rhythms of nature, ecosystems and all life. One of the pioneers of this area of interest is Georgescu-Roegen (1971). According to this author, mankind cannot continue fostering technological, economic, and ethical models, which are dangerous because they are aggressive and devastating for the Earth. This knowledge should encourage society to not only look for radical innovations in both scientific research and industrial applications, but also to foster change in the dominant consumer styles in our society.

What is lacking today is a sense of political will that would create and protect a new consciousness and a new environmental responsibility, as much scientific as civil, which is much too marginalized. According to

Latouche (2007) and Gare (2000), if it is true that there is no worst-case scenario for a society based on growth without growth, the workable alternative will be consuming less and more consciously to live better. The symptoms of having reached the limits of exploitation of the planet will affect the poorest countries in the southern hemisphere as well as future generations more acutely. Based on the abovementioned assumptions, this book aims to investigate the application of Degrowth theory in a resilient community in a Transition Town as a possible solution, a way to stem the global and current world economic and social crises.

The last thirty years have been characterized by a proliferation of alternative systems of this nature. Such phenomena are not limited to a specific geographical area. It is possible to find systems with names such as LETSystems, Time Banking Systems and Transition Towns based on non-monetary exchange, ethical purchasing groups, and transition towns in almost every country.

The Comparative Analysis with Two Hemispheres: Australia and Italy

This book is the natural result of my academic experience begun in Italy and then matured in Australia. The comparison, therefore, between Australia and Italy is a matter of great interest to me. The last time Australia's GDP did not grow the Soviet Union still existed and in Italy the political-judicial case known as "Tangentopoli" still did not break out. Since 1991, the Australian economy has not gone into recession—that is, it has not had two consecutive quarters with GDP down—even during the great global crisis of 2008, a record that has "no equal in any other country of the Organization for Economic Cooperation and Development" (Development Co-operation Report, 2018). For more than twenty-seven consecutive years, Australian GDP has grown at an average rate of 3.2%, resisting economic and political crises and, according to the latest figures published by the Australian Bureau of Statistics, in the second quarter of 2018 has grown again by 0, 9% compared to the previous quarter and 3.4% compared to last year.

The causes of this constant growth are different, but, undoubtedly, one of the contributions was given by the proximity to China, the main buyer of Australian raw materials, in particular iron and natural gas, of which Australia is very rich. However, the proximity to China, according to the Australian Bureau of Statistics (2018), has also helped another key sector of the country's economy, such as, tourism. Last year, Chinese tourists in Australia numbered 1.4 million, surpassing the number of tourists from New Zealand for the first time, and are expected to increase even more over the next ten years. According to the Reserve Bank of Australia (2019) GDP has grown again in 2019, reaching 3.5%. Despite this, however, there is no lack of worries and economists fear above all that the Australian economy is too dependent on the Chinese one, and that a major crisis in the Chinese economy—that is already suffering due to the duties imposed by the United States—can make the whole export system collapses.

For the "financial wizards", borrowing the terminology used by Latouche, Australia will be heading for recession, for the first time since 1991, in 2023. The excesses usually are the causes of a crisis, for instance, the average income of top managers, as Latouche reminds us (2006), that exceeds four hundred times the average income of salaried, the real estate assessments and so on. Perhaps it is not just a coincidence that the two most serious crises of capitalism coincide with the greatest inequality in income. On the other hand, in Italy, or rather, in Europe, nobody knows when a crisis will come. One of the closest indicators for the Italian economy is the slowdown of Germany, the largest economy within the European Union, and so far, one of the most dynamic in the world.

According to Nienaber (2019), during the second quarter of 2019, German GDP contracted by 0.1% compared to the first quarter of the year. As well as for Australia, even for the "old continent" one of the main causes of this slowdown is the bad period that international trade is experiencing, between the slowdown of the Chinese economy and the fears for an international trade wars, in particular the one that the United States has triggered against China and the one that could trigger against the European Union. Although in recent years there have been some attempts to increase domestic demand, the German economy is still largely based on exports: it therefore seems easy to understand that Germany is very attentive to the global market. The instability of Germany involves all European countries and in particular those that, such as Italy, are very tied to its economy.

Italy has already had a "slight" recession between the end of 2018 and the beginning of 2019, and other EU countries—such as Netherlands and Belgium, for example—which also produce components for German companies and are integrated into their production cycles. The signals of a new recession on a global scale are unequivocal. Signals such as the reversal of the yield curve on government bonds, one of the main indicators of an incoming recession. Other signals are coming from Japan and the United Kingdom. This one occupied by the complicated and delicate matter of Brexit. It seems definitely probable to venture that if it were to end with the infamous "no deal" it could bring further problems to the economic situation of the country and for the rest of Europe. Nevertheless, it could be summarized that this is the cyclical nature of capitalism.

For 30 "glorious years"—although Fourastie (1980), has argued that—approximately—every ten years the Western economies are disrupted, at national or global level, by rather critical shocks—the financial leverage has increased the rate of profit and in the meanwhile the salaries were reduced; even so the consumption was kept high enough through the indebtedness in its various forms. However, nowadays, as reminds us Sharma (2019), the world economy, in particular the one of the developed countries, has long been in difficulty.

However, according to the latest report of the International Monetary Fund (2019) the current world economic situation is delicate and growth estimates are not encouraging. On the other hand, monetary authorities, headed by the aforementioned financial wizard, they may decide to resort to less orthodox resources than usual. "Creative solutions" such as quantitative easing—or perhaps even more extreme—helicopter money, unorthodox measure imagined in 1948 by Friedman, that we will analyze in the following chapters.

Perhaps the aging of the population and the consequent reduction of the workforce, combined together with the instability of governments and a large-scale trade wars, are profoundly changing the conception of money. A spectre, maybe different and more terrifying from the one of the thirty glorious is imminent. With these, and certainly not encouraging scenarios, paraphrasing Walker and Salt (2006), more resilient is a system towards external disturbances, the more redundant it will be. Why not be ready (and resilient)?

According to Giovannini (2012), former Chief statistician of the Organisation for Economic Co-operation and Development (OECD), President of The Italian National Institute of Statistics and former Minister of Labour and Social Policies, due to the actions taken by local realities such as the Transition Town Movement and other national authorities, to research in relation to quality of life, to new paradigms such as Degrowth theory and, also, to the above mentioned research into the measurement of social progress by the Organization for Economic Cooperation and Development (OECD), a new movement is emerging aimed at measuring well-being or, trying to change the collective imaginary to develop a better understanding of the difference between well-being and well-having.

Is it possible to identify, within the Transition Town Movement, such as in the examples of this study i.e. Transition Darebin" and Fiorivano le viole, an indicator that could be an expression of ***buen vivir***? Will it be possible to liberate the concept of "happy" (Pallante, 2007) economic degrowth from its academic shackles and create a new useful tool for the emergence of *DemocraCity*?

The Future Is Now

This book examines a community in transition in Darebin in Victoria, precisely in Melbourne, Australia, and, a neighbourhood in transition located in Perugia, Italy, framed as a resilient, virtuous example of the application of Degrowth Theory and a workable example of inclusive democracy. The purpose of this book is to investigate the rise and spread of this phenomenon. Through the creativity of the social constructionist approach and use of the theoretical framework of Latouche's Degrowth and Resilience theories (Hopkins, 2009; Kendra & Wachtendorf, 2003; Lavanco & Novara 2012; Norris et al., 2008), this manuscript wants to provide a better understanding of the dynamics observed in transitional communities such as "Fiorivano le viole", in Perugia, and the dynamics this reality would face in everyday political activities.

The following questions have guided the path that drives this book: Does Latouche's theory of Degrowth and Resilience find a successful application in real life? Does Resilience have a role in creating better opportunities for communities who are facing difficult circumstances? Why does Degrowth frighten our collective imagination and why is resilience creating serenity within (virtuous) community? Transition towards what and how does local politics promote and support such a community?

The answers to these questions will contribute to the debate on the ecological approach that has historical, social, and economic roots. This manuscript aims to investigate what has been identified as a possible tool to achieve social innovation with the development of a social thermometer that can measure active citizenship, solidarity, communication and the planning skills of a community. In this study, Degrowth and Resilience are considered two sides of the same coin. The achievement of what the authors of Degrowth theory called GNH (Gross National Happiness) or *Buen vivír*, is only possible if members of society start developing values such as resilience of communities in order to embrace a workable, desirable major restructuring of our collective imagination.

This manuscript seeks to use as an example the journey of Transition Darebin" and Fiorivano le viole towards a delicate path that will bring (us to) a different and workable scenario for a sustainable future. Degrowth and resilience are the two necessary tools for the construction of what we call *DemocraCity*. After the failure of the homo *politicus* (Crouch, 2000) and the caducity of the *homo sustinens*—a category of human that Siebenhüner (2000) argues is characterized by a comprehensive understanding of sustainability cooperation and communication at the hands of homo *oeconomicus,* and after several attempts made by homo *ecologicus* (Becker, 2006) in order to rediscover and reinforce the balance between mankind and nature, the future lies in the hands of what we call homo *resaliens* and it is its responsibility to lead humans out of the consumer society.

As we mentioned above, virtuous examples of Degrowth, valuable example of the new civilization envisaged by Gare, have arisen all over the world. One example is the reality of Transition Darebin" , located in Melbourne. Another is Fiorivano le viole, located in Perugia. Both Transition Initiatives are used as a case studies in this research to demonstrate how communities based on principles of solidarity, mutual aid, virtuous networks and local resilience initiatives can, starting from the bottom, influence, change and ultimately bring about the end of old growth systems. The motto of Transition Darebin, "ready and resilient" was a source of inspiration for the creation of the aforementioned homo *resaliens*. Its etymology represents a novelty in the environmental literature. In addition, in order to further justify the contribution to knowledge, this book represents the first attempt to draw a map of the Transition Towns, virtuous communities, laboratories of alternatives paths located in Melbourne.

Furthermore, through the creativity of the social constructionist approach and use of the theoretical framework of Degrowth and Resilience theories this book wants to provide a better understanding of the dynamics observed in transitional communities and the complex dynamics these realities would face in everyday political activities.

From the "Sense of Community" to the "Civitas Maxima"

The methodological approach that drives this book includes thematic analysis based on McMillian and Chavis's approach called "Sense of Community". This approach includes the construction of a "social profile" of each case study community; the collection of survey data; the extensive use of semi-structured interviews, observation, and the collection of "found stories" collected on field. To this end, qualitative methods are used to provide a better understanding of the subject and the possible outcomes for Australian and Italian communities. Relevant national policies and programs on building resilient communities in Australia and Italy are analysed and compared in order to understand whether nation states are playing a role in building new resilient communities.

In addition, secondary data and statistics are gathered from governmental websites and from international organizations and NGOs. Academic research using cases like Transition Initiative such as Transition Darebin, to discuss Degrowth theory, the application of resilience for human communities to create an alternative way of thinking to overcome the current economic model, is not present in the existing literature.

In addition, there are no scientific works that trace the map of the Transition Towns that exist in Melbourne, of which Transition Darebin" (TD) is just one of many examples. On the other hand, although there is an important scientific production which helps to understand the phenomenon of the Transition Initiatives in Italy, a study that connects and compares two such different and distant countries, represents a novelty in the broad area of the Degrowth Theory on a larger scale.

References

Alexander, S. (2012). Peak oil and the twilight of growth. *Alternative Law Journal, 37*(2), 86–90.

Andreoni, V., & Galmarini, S. (2014). How to increase well-being in a context of degrowth. *Futures, 55*, 78–89.

Australian Bureau of Statistics. (2018). Annual Report, 2018–19. https://www.abs.gov.au

Barret, G. (2012). Explorations in the economics of aging. *Economic Record, 88*(281), 290–292.

Becker, C. (2006). The human actor in ecological economics: Philosophical approach and research perspectives. *Ecological Economics, 60*, 17–23.

Beckerman, W. (1974). *In defence of economic growth*. J. Cape.
Belpomme, D. (2004). *Ces maladies crées par l'homme*. Albin Michel.
Besset, J. P. (2006). *La scelta difficile. Come salvarsi dal progresso senza essere reazionari*. Dedalo.
Bloemmen, M., Bobulescu, R., Le Tuyen, N., & Vitari, C. (2015). Microeconomic degrowth: The case of community supported agriculture. *Ecological Economics, 112*, 110–115.
Bornestein, J. (2013). What to save? The normative dilemmas of resilience. *Centre for Human and Nature, 6*(3), 7.
Chochet, Y. (2009). *Antimanuel d'écologie*. Breàl – Parution.
Crouch, C. (2000). *Coping with post democracy*. Fabian Pamphlets.
Daly, H. E. (2010). *Ecological economics: Principles and applications*. Island Press.
Development Co-operation Report. (2018). *Joining forces to leave no one behind*. https://www.oecd-ilibrary.org/
Fourastie, J. (1980). *Les Trente glorieuses ou la Révolution invisible de 1946 à 1975. Le Livre De Poche Things*. New York: North Point Press.
Hopkins, R. (2009). *Manuale pratico della transizione: Dalla dipendenza dal petrolio alla forza delle comunità locali*. Arianna, Bologna.
Gare, A. (2000). Aleksandr Bogdanov's history, sociology and philosophy of science. *Studies in History and Philosophy of Science, Part A, 31*(2), 231–248.
Georgescu-Roegen, N. (1971). *The entropy law and the economic process*. Harvard University Press.
Giovannini, E. (2012). Indagine conoscitiva sull'individuazione di indicatori di misurazione del benessere ulteriori rispetto al PIL. *Audizione del Presidente dell'Istituto nazionale di statistica Enrico Giovannini, Commissione V "Bilancio, tesoro e programmazione" della Camera dei Deputati*.
Godbout, J. T. (1998). *Il linguaggio del dono*. Bollati Boringhieri.
IMF. (2019). Annual Report 2019. https://www.imf.org
Jackson, T. (2009). *Prosperity without growth. Economics for a finite planet*. Earthscan.
Jackson, T. (2011). *Prosperity without growth: Economics for a finite planet*. Earthscan.
Kallis, G. (2011). In defence of degrowth. *Ecological Economics, 70*, 873–880.
Kallis, G., Kerschner, C., & Martinez-Alier, J. (2009). The economics of degrowth. *Ecological Economics, 84*, 172–180.
Kallis, G., Martinez-Alier, J., Schneider, F. (2010). Growth, recession or degrowth for sustainability and equity? *Journal of Cleaner Production*.
Kendra, J., & Wachtendorf, T. (2003). *Disasters: Elements of resilience after the World Trade Center disaster: Reconstituting New York City's Emergency Operations Centre*.
Kerschner, C. (2010). Economic de-growth vs. steady-state economy. *Journal of Cleaner Production, 18*, 544–551.

Kimhi, S., & Shamai, M. (2004). Community resilience and the impact of stress: Adult response to Israel's withdrawal from Lebanon. *IEEE Journal of Community Psychology, 32*(4), 439–451.
Latouche, S. (2006). *The globe downshifted*. Le monde diplomatique.
Latouche, S. (2007). *La scommessa della decrescita*. Feltrinelli.
Lavanco, G., & Novara, C. (2012). *Elementi di psicologia di comunità. Progettare, attuare e partecipare il cambiamento sociale*. McGraw-Hill Education.
Muraca, B., & Schmelzer, M. (2017). Sustainable degrowth: Historical roots of the search for alternatives to growth in three regions. In I. Borowy & M. Schmelzer (Eds.), *History of the future of economic growth: Historical roots of current debates on sustainable degrowth* (pp. 174–198). Routledge, London.
Nienaber, M. (2019). *Germany's Chinese growth engine stalls*. https://www.Reuters.com
Norris, F. H., Stevens, S. P., Pfefferbaum, B., Wyche, K. F., & Pfefferbaum, R. L. (2008). Community resilience as a metaphor, theory, set of capacities, and strategy for disaster readiness. *IEE American Journal of Community Psychology, 41*.
Pallante, M. (2007). *La decrescita felice. La qualità della vita non dipende dal pil*. Editori Riuniti.
Reserve Bank of Australia. (2019). Annual Report 2019. https://www.rba.gov.au
Sharma, R. (2019). *Small is the new big thing economic forecasters got the 2010s wrong. What will the next decade bring?* https://Nytimes.com
Siebenhüner, B. (2000). Homo sustinens—Towards a new conception of humans for the science of sustainability. *Ecological Economics, 32,* 15–25.
Trainer, T. (2012). De-growth: Do you realise what it means? *Futures, 44*(6), 590–599.
Walker, B., & Salt, D. (2006). *Resilience thinking: Sustaining ecosystems and people in a changing world* (1st ed.). Island Press.

CHAPTER 2

The Rise of a Virtuous Consciousness

The intention of this chapter is to review the existing literature that led to, among other things, the creation and development of Serge Latouche's Degrowth theory and its impact on global communities. This chapter will also seek to identify possible and achievable approaches discussed in the literature of the broad area of Degrowth and resilience that still require examination and comparison for a better understanding of this theoretical framework. Moreover, this chapter provides an overview of the literature of Degrowth Theory: resilience and the transition towards a resilient community.

The analysis of the literature on Ecologic Economy, is a crucial element in this study. To suggest an alternative narrative in which characteristics of three actors of a holistic economic will be combined and compared to "*homo oeconomicus*": "*homo sustinens*" (Siebenhüner, 2000), "*homo politicus*" (Crouch, 2000), "*homo ecologicus*" (Becker, 2006) in order to find within *homo "resaliens"* the concretion of the values such as altruism, mutual aid, reciprocity, communication, human values, and, as an imperative, responsibility for future generations.

The phenomena of environmental crises and the exploitation of natural resources have been studied and debated for more than a century. A combination of different scholarly views and disciplines in tandem with a differing economic and ecological analysis have provided a new way of thinking—focusing not only on empirical evidence and the causes

M. Carcea, *The Road Towards DemocraCity*,
https://doi.org/10.1007/978-981-16-3505-2_2

of the aforementioned phenomena, but also on the solutions, alternatives, and possible ways in order to overturn our current (and dominant) socio-economic system that jeopardizes the survival of current and future generations.

The Beginning of a New Discipline

The word "ecology" was coined in 1866 by the German biologist Ernst Haeckel, who, in revealing Darwin's findings had suggested the need for an autonomous discipline in order to describe the influence that the environment exerts on living beings. Such a new discipline would have to describe both the exchange of matter and energy between living beings and the atmosphere water, oceans, the ground, both exchanges of living beings between them, joined by food chains. According to Odum (1971), Commoner (1986), Conti (1987), and Arrow et. al. (1995), economists, not biologists were the ones who raised issues about the relationship between humankind and the environment.

The research conducted by these scholars contributed directly to the realization of the foundation of the environmental issue of the relationship between mankind and the resources it needs to live. It was their intuition that led classical economists to theorize that demographic growth would lead to a decrease in the average productivity of cultivated land; the pioneering study of Malthus stressed reflection on the origin of population growth and the progress of the curve which expresses the trend. Other important "classical economics" scholars of the late eighteenth century, fathers of the mercantilist economy, such as Adam Smith, David Ricardo and John S. Mill also showed, an awareness of environmental issues. They emphasized the relation between the ability—and the opportunity—of economic development to allow continuous improvement of material living conditions in the long run.

Latouche (2006), and later Jackson (2011), remind us that it was the industrial revolution that ushered in economic development. At the same time, these developments raised questions of what economic development might mean or whether it would or could last. An answer, however uncertain, given by those scholars is to be found in the law of diminishing returns which postulates that there is a point at which the level of benefit gained from production is less than the level of capital or energy invested. It seems perfectly clear that the classical approach, according to Piketty (2013), and Esposito et al. (2018), despite the fact that it had a purely agricultural application (due to the characteristics of the economic system of the time), is closely linked to the problem of scarcity of natural resources; resources that population growth would help to decrease dramatically.

According to Cacciari (2006), putting a stop to economic growth seems to be an answer. The contribution of Malthus and Ricardo was the introduction and explication of the concept of soil load capacity. The finite nature of natural resources and the limited possibilities of using them represented the limit for both demographic and economic growth. Despite the limits of classical economics, constraint due to scarcity of resources are a cornerstone in the studies of the two authors for a proper interpretation of development. According to Nelissen et al. (1997), the studies are mainly mono-disciplinary, are in need of a policy reference, often not do not include any political statements and are mainly of a scientific nature. However, according to this author, they are the cornerstones of modern environmental studies in the sense that the issues and statements they introduce are still topics in the current policy environment.

The Malthus paradigm was extended to include a new, "Neo-Malthusian" variable: the variable of "misery" caused by overpopulation. The significance of such a variable is quite clear: the degradation of environmental quality due to pollution, related to the progressive expansion of the economic processing model, has created the search for higher levels of well-being pursued by an increasing number of individuals. The English economist Arthur C. Pigou extends the Malthusian approach synthesized in the *An Essay on the Principle of Population* (1798) with his critique of GDP, perhaps the most important issue of Degrowth theory. According to Nelissen et al. (1997), Stiglitz (2010), and Mattauch et al. (2018), Pigou questions whether Gross National Product is still an effective tool to measure welfare and, whether/how this measure includes negative external effects, such as the depletion of natural resources, the influence of pollution, and contamination to the public health. Kallis et al. (2010), synthesized, in order to define the Degrowth, what Pigou preconized "*an equitable down-scaling of production and consumption that increases human well-being and enhances ecological conditions at the local and global level, in the short and long term*" (p. 512).

In 1920, in his book *The Economics of Welfare*, Pigou stressed that there were cases in which the market mechanism did not lead to an efficient allocation of resources because of what economics calls "negative externalities". Negative externalities occur when the consumption or production of good causes a harmful effect to a third party, without a monetary compensation being made, i.e. without the market price system being taken into consideration. Pigou's theoretical approach tends to embrace a socio-economic vision diametrically opposed to the agricultural vision set by Ricardo and Malthus. The very fast industrialization of the masses has demonstrated how the benefits produced by it did not extend to the whole of society. In Pigou's view, the benefits of rapid industrialization did not extend to the whole of society.

The capitalist system has produced exploitation, unfavourable conditions for workers, child labour (issues addressed by Marx as well). Such themes led Pigou to ponder the relationship between social and private net product. "The essence of the matter is that a person A, in the course of rendering some service, for which payment is made, to a second person B, incidentally, also renders services or disservices to other persons (not producers of like services) of such a sort that payment cannot be exacted from the benefited parties or compensation enforced on behalf of the injured parties. [...] we may set out first a number of instances in which marginal private net product falls short of marginal social net product, because incidental services are performed to third parties from whom it is technically difficult to exact payment. [...] Corresponding to the above investments in which marginal private net product falls short of marginal social net product, there are a number of others, in which, owing to the technical difficulty of enforcing compensation for incidental disservices, marginal private net product is greater than marginal social net product" (Pigou, 1920, II.IX.10).

In this extract, we come across the work of K. William Kapp and his theory of the *social costs* of economic activities. Based on this concept, Pigou proposed the solution of introducing corrective taxes, to be paid by industry (Pigouvian Tax) in order to rebalance the private costs of industry and the social costs of the entire community. Although, as set out above, Kapp extended the framework to the social sphere, the German economist had to engage with and move beyond the limits of Neoclassical economics. According to the Neoclassical approach, it is possible to evaluate and contain the impact of the degradation of social life and of the environment through a system price.

To resolve this issue Kapp argues that the system of dynamic relationships that connects the natural environment economy cannot be analysed on the basis of the price system. The German author emphasized that the relationship system must include physical parameters after selecting the relevant elements by virtue of their use value. Only through this process, according to Kapp (cited in Calafati, 1997), can we get a clear understanding regarding the system and the relationship of interdependence between the economic and the natural environment process.

Matthey (2010), Graeber (2011), and Piketty (2013), argue that the end of World War II can be identified as the beginning of the process of economic growth as all Western countries at that time had as their main objective the spread of economic well-being and democracy. In addition, the dynamics of the world's political climate were shaped by the Cold War and the divisions between the Western and the Soviet world. Industrial power had played a massive role in the war debate, and the development of productive capacity was considered the main key needed against the Soviet threat. Thus, it became an article of faith that Western ideological superiority was demonstrated by rising GDP growth. GDP growth became a "sacred" phenomenon that, in the late 1940s and almost all of the 1950s, was immune from any criticism. Any issue regarding the environment was simply not relevant.

According to Minear (2011) and Brougher (2015) a reversal of direction took place in the mid-1950s. Nuclear testing due to its effects on the environment became a contested issue. Emblematic in this regard was the episode of the "Lucky Dragon" that occurred in 1954 in the Pacific. The episode recalled the nuclear devastation of Hiroshima and Nagasaki. Very soon the world became aware that nuclear testing gave rise to the formation of long-term radioactive isotopes that, through deposits in the soil and the sea, were absorbed by the environment and by all living creatures, including mankind.

Such awareness shifted scientists and public opinion and introduced a new sensitivity, succinctly expressed in the words of Albert Schweitzer (Nobel Peace Prize in 1952), in the introduction of Carson's *Silent Spring* (1962): "Man has lost the ability to anticipate and prevent; he will end up destroying the Earth". Members of the *intellegentia*, inspired by the strength of Schweitzer's and Carson's words, were inspired to look for new theories, new methods, in order to find new ideas and tools that could address environmental pollution. This is a clear demonstration of how the growing influence of the environmental movement was instrumental in changing the foundations of society as a whole. Carson (1962) evidenced the environmental consciousness which had risen to a new level of sensitivity. The American biologist exposed the effects and damage to health caused by prolonged exposure to chemicals.

It was a turning point, both in the scientific community and in public opinion in terms of the relations of interdependence between economics, health, and the environment. Another important contribution to raising awareness was given by Leslie Reid (1962). For the first time an ecocentric approach was adopted and unlike in the anthropocentric one, man and nature were accorded parity. Hardin (1968) emphasized the problem of an increasing population with rising material consumption and the impossibility of their unlimited growth due to the finite nature of natural resources. The matter of the growing population is also raised by P. and A. Ehrlich (1969). Biology was being increasingly woven into the interpretation of society.

Also, E. P. Odum (1969), identified two types of ecosystems: the first is a young one, in which production is high; at this point, the ecosystem is characterized by growth, the number of species being more relevant than the quality. These types of ecosystems are interesting for mankind as they can provide us with food and resources that we need. In contrast, we have the highly stable mature ecosystems, in which the quality of species is more relevant than the quantity. These ecosystems do not have a significant role in production, as this level is low; they provide us, however, with other features relevant to humankind. The three elements that characterize the young ecosystem, i.e. production, quantity, and growth, are replaced by new elements that define the mature ecosystem: protection, stability, and quality. This approach is relevant because it stressed the fact that the stage contemporary societies exist in is definitely not one situated in the young ecosystem and, for this very matter, it is absolutely necessary to change the perspective of our ecosystem. A new era was emerging in which changes were affecting education, planning, environmental policies, and social institutions. Such change, regarding the environment, was not called for exclusively by biologists. According to Calafati (1997), Fisher and Ponniah (2003), Latouche (2009), and Anguelovski (2015) environmental economics is a field that occupies a primary place in economic thinking. It is widely recognized that in the *mare magnum* of disciplines, Economics was one of the first that has faced the study of the phenomenon of environmental degradation. K. William Kapp (1971) offers a fundamental contribution to the construction of an environmental economy, with concise criticisms of the orthodox economy, more precisely the environment and development economy, until he comes to the definition, both theoretical and practical, of an alternative approach. Kapp's contribution can be seen as that of a pioneer, futuristic in his criticism of the canonical economy and perhaps, this is the reason why he has not received the proper recognition from the scientific community.

Only in recent years, according to Heidenreich (1998),Adkisson (2009), and Valentinov and Chatalova (2013), Kapp received a greater and well-deserved recognition. The cornerstone of his way of thinking is the representation of the economic system as an open system, in clear contrast to the closed representation on which the modern concept of economics is based. Starting from the assumption that the traditional economic system ignores the dynamics of exchange between the social and natural environment, Kapp (1971) emphasized how the economic process is actually interconnected with a whole set of elements of the physical and social world, that it is a much wider process than orthodox economists seem to feel.

Such interconnection is the economic process that changes the value of use of elements of the natural investigation through multiple implications. According to Daly (1992), environmental degradation can be defined, to justify the investigation from his multidisciplinary perspective, as a reduction (or loss) of the value of the elements of the natural environment. of Kapp's perspective makes it clear that environment degradation is part of the category of social costs. In particular, Kapp (1971) focuses on the definition of "waste" attributing this to consequence of the capitalist economy and high society behaviour.

As set out above, according to Adkisson (2009), Kapp's view (1971) of social cost represents the cornerstone of his work: from the costs of air and water pollution, through the intensive exploitation of resources, to the investigation of the damages to the social and physical environment such as occupational diseases, accidents, exploitation of female, and child labour. Moreover, according to Heidenreich (2016), Kapp provides a detailed analysis of technology, criticism of the market economy, concentrating his efforts on structural and cyclical unemployment, social costs linked to excess production capacity in the transport and retail sector, on the bad location of industrial complexes, the cost imposed by the patent system, and last but not least the programmed obsolescence of goods.

Berger (2008) reminds us that the character of Kapp's work is not purely economic; Kapp's work highlights themes of seminal importance. They include reference to the systemic interdependencies between the economic and the socio-environmental world, and criticism of optimism and economic efficiency as a method of social assessment. The work, according to Swaney and Evers (1989), also includes the role of technology and institutions, of essential goods and non-essential goods, of price theory and value theory, and the objectification of certain needs, which can only be recognized by their absolute meaning and rediscovery of certain values. Two ethical norms characterize the work of Kapp: we should not jeopardize humans living on Earth; and we have to minimize human suffering.

Cowboys and Astronauts

As has been shown, the seeds of the future Degrowth theory were already *in nuce* by the late '50s. In 1967, with *The Historical Roots of our Ecological Crisis*, L. White identifies that the way of thinking and the development of science and technology in the Western world over the last two millennia, was heavily influenced by the Christian idea of humankind as dominant and sovereign on the natural world. However, Boulding (1966) questions dominant economic theory by questioning why Gross Domestic Product is the sole index of economic and social welfare:

> Systems may be open or closed in respect to a number of classes of inputs and outputs. Three important classes are matter, energy, and information. The present world economy is open in regard to all three. We can think of the world economy or "econosphere" as a subset of the "world set", which is the set of all objects of possible discourse in the world. We then think of the state of the econosphere at any one moment as being the total capital stock, that is, the set of all objects, people, organisations, and so on, which are interesting from the point of view of the system of exchange. This total stock of capital is clearly an open system in the sense that it has inputs and outputs. (Boulding, 1966, p. 3)

Input, since the economic system receives flows of raw materials and natural resources from the natural system; *output*, because the natural system receives waste streams from the production and consumption activities. A seminal work was *The Economics of the Coming Spaceship Earth* (1966), in which the English economist stressed the need to move from the cowboy economy to the spaceman economy. The first economy, according to Boulding (1966), is characterized by open and unlimited spaces, by violent behaviours which are directly related to exploitation. In the second economy the earth has become a single spaceship, without unlimited reservoirs of anything, either for extraction or for pollution, and in which, therefore, man must find his place in a cyclical ecological system which is capable of continuous reproduction of material form even though it cannot escape having inputs of energy.

According to Höhler (2007), Helmreich (2011), and Victor (2014), the difference between the two types of economy becomes most apparent in the approach towards consumption. In the cowboy economy, consumption is regarded as a good thing and production likewise. The success of the economy is measured by the amount of throughput from the "factors of production", a part of which, at any rate, is extracted from the reservoirs of raw materials and noneconomic objects, and another part of which is output into the reservoirs of pollution. If there are infinite reservoirs from which material can be obtained and into which effluvia can be deposited, then the throughput is at least a plausible measure of the success of the economy. Gross National Product is a rough measure of this total throughput.

In contrast, Boulding argues, in the Economics of the coming spaceship Earth, throughput is by no means a desideratum, and is indeed to be regarded as something to be minimized rather than maximized. The essential measure of the success of the economy, as Deese (2009) reminds us, is not production and consumption at all, but the nature, extent, quality, and complexity of total capital stock, including the state of the human bodies and minds included in the system. In the spaceman economy, what we are primarily concerned with is stock with a lessened throughput (that is, less production and consumption) which is clearly a gain.

The English economist was one of the first to recognize how to evaluate production and consumption activities in a negative as opposed to a positive manner. This may seem odd to conventional economists obsessed with income streams. It seems to be clear that during the 1960s the path that lead to the criticism of economic growth began to take form. *The Cost of Economic Growth* (1967) is another cornerstone of the literature that empowered such multidisciplinary work. Mishan continued and reinforced the work of Boulding establishing explicitly the subject of environmental criticism, i.e. economic growth.

Mishan's study, as Jaffe et al. (2005) remind us, was based on simple and measurable economic parameters, such as employment, income level, public debt, and exchange rates. However, according to Andrew (2008), by concentrating on these easily measurable economic parameters, other relevant economic issues are overlooked and neglected. Andrew (2008), argues that Mishan brought the measurement of economic value, in particular when environmental issues are at stake to the core of the economic debate. Moving beyond the Pigouvian approach of negative externalities (costs not internalized in the market price), Mishan argued that market prices do not truly reflect the total economic scarcity which it had to give up in order to produce a good, i.e. production cost does not include environmental costs.

Although he recognized the necessity to include the environmental costs in the price of products, Mishan, according to Vatn (1998), struggled to quantify environmental destruction in economic terms. Mishan thus fails to go beyond the fundamental difficulties of the environmental economics of his time. However, his contribution was a cornerstone for the literature because, according to Dahlman (1979), Nelissen et al. (1997), Helmsing (2001), Bianchi (2011), Mishan argues that is not certain that an increase in the level of production results in an increase in welfare, as is assumed by traditional economists. If an increase in production creates a more than proportional increase in negative external effects, the increase in production will result in a decrease of welfare. Welfare could be increased by lowering the level of production, as this would lower the environmental costs so sharply that an increase in welfare would be the result.

The 1970s, the Idea of "Growthmania" and the Club of Rome

As we get closer to the golden age of the ecological economics of the 1970s, we also get closer to what has been termed as *growthmania*. This concept, developed by economist Colin Clark, was also the creator of the gross national product indicator (GNP). The Australian economist was one of the main critics of the ecological objections, according to whom he stated that the Earth could easily feed forty billion people. According to Latouche (2009), Sen (2011), and Piketty (2013), these were the years of uncontrolled development, achievable "solely" by producing and consuming more goods, exploiting more and more forests, water and the scarce resources of the earth.

The Tektology of A. Bogdanov (1921), in which Nature represents the prime engine, and basis of the System Theory and Universal Science of Structures, had been completely forgotten. Nature is part of every single human process (from growing complexity to social experience) but, in addition, Nature is within those entities considered as elementary in the past (Gare, 2000). For *growthmania*, the lesson as set out by the Marxist Russian theorist Bogdanov, i.e. the "not superficial" faith in the organizational human and natural action, was no longer important. The uncontrolled violence perpetrated against nature and its resources, is also represented in the growing inequality between the exploiters and the exploited.

In response to this terrifying scenario, Barry Commoner published *The Closing Circle* (1972), acknowledged as one of the cornerstones of ecological thinking, and perhaps forerunner of the emerging theme of Political Ecology. The idea was already in the title: Nature works with closed cycles: the water cycle, oxygen, carbon, nitrogen, phosphorus. Natural transformations are powered by the energy of the Sun. Matter always stays within the cycle and is reused. Nature knows no waste: chemical substances extracted from air, from water and soil, are returned to circulation and again become raw materials for other natural cycles. Environmental degradation and pollution lead to the collapse of the natural cycles.

Closed processes cycles shift to become open, and there is no longer a balance between what is extracted from and what is returned to the reserves of natural resources. Waste increases in such quantities that Nature can no longer assimilate the waste products. Salvation lies in the closure of natural cycles, through immediate action in the political, scientific, and technological fields. *The Closing circle* was an invitation to review the prevailing models of consumption. In t modern industrial societies, companies survive only if they produce more goods at the lowest possible cost and this involves further exploitation of nature and more pollution.

According to Conti (1987), the Barry Commoner approach does not focus on the amount of the production processes that degrade the environment, but the way goods are produced. The issue he mentions is thus the polluting production methods, against which citizens have a duty to fight in order to protect and defend their fellow workers as well as the environment. Needless to say, Commoner' contemporaries view his approach as including subversive and provocative content.

1972 was also the year in which a circle of people much less radical than Commoner, the *Club of Rome*, commissioned J. W Forester and D. L. Meadows, two American scholars from MIT, to produce a report that would serve to relate the reciprocal influences of world population, agricultural, and industrial production, the exploitation of natural resources and pollution. The book *The Limits to Growth* presented in graphic form the principles as posited by Commoner (Turner, 2008). It demonstrated that a simultaneous increase in population and agricultural and industrial production would lead to an equivalent depletion of natural resources and environmental degradation to the point of causing famine, war and disease and thus decimate a good portion of the world population.

Limits to Growth called for a courageous and profound renewal of society through technological, cultural and institutional activities designed to prevent the ecological footprint exceeding the carrying capacity of the planet. Supported by mathematical models that consist of five basic variables: population, food production, pollution, natural resources and industrial capital and meticulous analysis of the models, such an extrapolation to future performance is accomplished in several scenarios. The conclusions drawn by the MIT scholars highlighted the physical impossibility of maintaining the period of growth without paying a terrible price for the next 100–150 years. However, there were also dissenting voices. Robert Solow (1974) who believes in the power of technology rejected the alarming report commissioned by the Club of Rome regarding the world ecological sustainability.

The book promoted by the Club of Rome, ended with an appeal to stop population growth, production and consumption. However, according to Golub and Townsend (1977), LaRouche (1983) and Bailey (in his "infamous" article written for Forbes in 1989, the Club of Aurelio Peccei contribution lacked the causes of the degradation of nature and was hesitant regarding the remedies. Herman Daly, one of the founders of the field of ecological economics and a leading critic of neoclassical growth theory, defines the concept of the steady state economy inspired by John Stuart Mill. The English philosopher in his "*Principles of Political Economy of the Stationary State*" (1911), argued that the logical conclusion of uncontrolled growth is the destruction of the environment and such unending growth will lead us to a significant reduction of our welfare.

Daly took Mill's lesson and designed the "*balance biophysical economics and moral growth*" (Daly, 1981), in which he stressed that an economy with constant stocks of people and artefacts, maintained at some desired, sufficient levels by low rates of maintenance "throughput", that is, by the lowest feasible flows of matter and energy from the first stage of production to the last stage of consumption. From the Seventies onwards, the literature based on degrowth has been enhanced by the contribution of Nicholas Georgescu-Roegen (former student of Joseph Schumpeter and future mentor of Herman Daly).

Nicholas Georgescu-Roegen was not merely the mentor of H. Daly but was also acknowledged as one of the pioneers of the Steady State Economy during the eighties. His was one of the first attempts to formulate a new paradigm and a new alternative in opposition to dominant neoclassical economics. The assertion of the impossibility of infinite growth in a finite world becomes the central economic analysis for the very first time with Georgescu-Roegen (1971). He introduced the physical laws of thermodynamics and entropy into the economic discipline, also extending them to matter as well as energy and implementing the imaginary substitution of the "*mechanical pendulum*" with the "*hourglass thermodynamics.*".

Since the Seventies, he has continuously stressed the need to replace traditional neoclassical economic theory with the theory of bioeconomy. The Romanian economist highlighted how the traditional economy ignores the law of entropy, i.e. the irreversibility of the transformations of energy and matter. Our economic "*human growth*" collides with the limits of the biosphere because the regenerative capacity of the planet is no longer able to satisfy demand. Man transforms the resources into waste faster than nature is able to transform such waste into new resources. Georgescu-Roegen advocated for a revolution of the economy to turn it into a bioeconomy, a dramatical resetting of the patterns of production and consumption: our "*salvation*" can take place by a *decrease* of production and global consumption; our salvation can occur only through a change in the human economy.

During the 1970s, E. Goldsmith published *A Blueprint for Survival* (1972). The Anglo-French environmentalist went beyond a critique of the dominant development model and its impact on nature. He framed his work into a political perspective; he tried to suggest a political response by indicating alternative solutions that should be adopted: our system of production and consumption must be in harmony with natural ecosystems in order for sustainability to not only be a target for the present, but to remain one for future generations also. According to Goldsmith, the aim of the blueprint is to convince governments, labour unions, and citizens of the need to change society itself.

Goldsmith formulated the main principles of a stable society that can sustain itself infinitely without jeopardizing the well-being of its members. It should involve a minimum destruction of ecological processes, a maximum conservation of matter and energy, a stable population without growth elements that can add pressure to the environment, and a social system that ensures that the three conditions above are not a restriction to the individual. In order to achieve such a transformation, he suggested seven operations:

- Reduce the environmental destruction through the development and implementation of new technologies in the production process.
- A period of stabilization, required to pause the exponential growth of production, and the processes of exhaustion and destruction of resources.
- A process where the elements of the most damaging trends are analysed and replaced by alternative technologies that in the short term are less negative and destructive compared to the first one.
- It would require a systemic substitution that introduces natural and self-regulating technologies instead of the previous technological alternatives, whose biggest difference is the long-term sustainability.
- The development, promotion and application of technologies that preserve energy and matter, designed for the relatively closed economic communities, which may cause only a minimal destruction of the ecological processes.
- A decentralization at all levels, both the political power and of the economic process, in order to achieve autonomous communities able to self-regulate and self-sustaining.
- Education at the community level (Goldsmith, 1972).

This systemic substitution would allow a reduction that applies to the transport of matter and energy and an easier control over the flows of goods and waste, and furthermore a control on resource depletion and pollution; community-level education. Goldsmith can be considered one of the main scholarly voices to criticize "insane" development, pointing out the need for radical change in society as well as in the economy, responding to his detractors and political counterparts through the conceptualization of theories that support an anarchic utopian socialism (and therefore considered unattainable).

In 1970 *Earth Day*, a day dedicated to our planet, was celebrated for the very first time to strongly affirm the need to safeguard the conditions for the survival of our planet. Tens of thousands of people, who came out into the streets, represented a signal that the urgency of ecological problems, pollution damage, and disparate use of land resources was not only felt by a minority of "dreamers" or fighting scientist, but was by now becoming a matter that had gained traction across all levels of society and involved concerted action.

Another work of great significance was the work of Lovelock: *Gaia. A New Look at Life on Earth*. According to the English chemist, the Earth is a "*quasi-living system*" with self-stabilizing capacity; the title refers to Gaia, the Greek goddess of the earth. From the point of view of this research, Lovelock's ideas can be evaluated as a positive instrument to combat negative ideas which were rather common in that decade. On the other hand, it has to be argued that his ideas as set out in the book seem to lack any (traditional) scientific basis.

Lovelock's ideas should be placed in the context of the debate at the end of the sixties and the seventies in which the attitude towards environmental pollution and how to deal with it was at the core of the social debate. Later on, in the eighties, environmental policies became much more accepted as "normal" phenomena in society, which reduced the relevance of Lovelock's ideas" (Nelissen et al., 1997).

Development and Its Side Effects

The scientific approach to environmental issue begins to be questioned by a solid narrative in the eighties. These years are characterized by many "trigger events" or environmental crises, such as the nuclear disaster in Three Mile Island in 1979, the Bhopal gas tragedy in 1984, Chernobyl in 1986, the Exxon Valdez oil spill in 1989. However, on the other hand the world, ruled by Regan and Thatcher, blindly followed the market mechanism. According to Nelissen et al. (1997), such controversial figures overlooked the fact that the market can only realize an optimal use and efficiency of priced inputs in the production process, which implies that the unpriced input of natural and environmental resources were not given full attention. They did not really have "a close relationship" with the environment. Many people were confronted with an economic recession: their jobs were on the line and they were forced to concentrate on "bread and butter" issues. In other words, the general, social and political climate was not very favourable for the environment. This was such a fertile ground to focus on privatization, liberalization and deregulation policies, and on the other hand to focus public attention and concern on employment and income problems.

That environmental awareness of the seventies was put aside, but the reality was also different. It was not that environmental awareness had been abandoned but more that there was a social and political infrastructure that did not encourage a proper green way of thinking. According to Redclift (2002), Latouche (2007, 2013), despite the relatively unfavourable social and political climate, the 1980s were not without value to the environment and environmental policy. Environmental issues were permanently placed, according to Jones (2010) and Glasson (2012) on the political and policy agenda of many Western countries. In many countries, a department or bureau was created to develop environmental policy. These institutions were an important counterbalance against the ruling conservative ideology of the 1980s. The degree of institutionalization of environmental policy was such that it could no longer be ignored (Nelissen et al., 1997).

The foundations were laid for education in an institutionalized environment. This increased the number of people directly or indirectly involved in a new way of thinking. It was an important mechanism for the defence and promotion of an ecological philosophy, and moreover the environmental policy landscape showed signs of growth, due to the introduction of legislation and environmental regulations to contain air, water and soil pollution, according to the "command and control" principle. In a climate of unrestrained deregulation, which had popularized and shared the imperative of reducing laws, especially in the economic field, such environmental norms and restrictions were naturally among the first to be criticized and attacked because of their incompatibility in the global competitive scenario.

According to Nelissen et al. (1997), such a scenario encouraged the development of more modern environmental instruments geared primarily towards incentives rather than restriction. Emphasis was put on the behaviour of the regulatory effects of market promotion policies using instruments in the form of taxes and subsidies. In the course of the eighties, Jackson (2011), Kempf (2010), and Piketty (2013), remind us that a number of financial incentives were enforced to encourage polluters to act in a more environmentally responsible way. These incentives came in many forms, for example, subsidies to stimulate environmentally correct behaviour, the manufacturing of environmentally correct products and production, but also taxes to reduce environmentally burdening activities. The business community was very much against the taxes that were being imposed, but in many countries, government acted with vigour and left business no choice.

It was clear that a new economic paradigm was emerging, the Ecological Economy. The defeat of the thesis of the 70s was definite. In the early 1970s, according to Daily et al. (1997), Pimentel et al. (1997), even the neoclassical authors had dealt with the relationship between the economy and the environment, reaching a distinction between the environmental economy and the economy of natural resources. The first one, according to Burke et al. (2015), was aimed at studying the interactions between economic and environmental system on the side of the effects (pollution) that economic activities exerted on the natural system. The object was therefore the management of common property resources.

The second, on the other hand, Kallis argues (2017) had its goal in the good exploitation of exhaustible and renewable natural resources. Despite this distinction and clear disciplinary autonomy, the analysis remains exclusively at the microeconomic level. The environmental economy (later on Ecological Economy) remains, according to Jackson and Webster (2016), strongly anchored in the neoclassical economy and its methodological approach which interprets environmental degradation as the set of environmental externalities generated by the economic process, and in which environmental externalities are seen as the consequence of a market failure. When the market fails, according to Jaffe et al. (2005) the economic process generates environmental externality, that is, a non-negotiated (and non-compensated) use of natural capital elements employed in the process of consumption or production of another agent.

The failure lies in the impossibility to reach an efficient allocation of resources, and therefore, the maximum collective welfare that the neoclassical environmental economy responds through or a monetary valuation of the environmental good considered, or the choice of the best tool to make the missing compensation. The goal, to achieve "Pareto efficiency", is clear. According to the market mechanism, characterized by agents whose rational decisions are based on calculations in monetary or one-dimensional terms, the solution is the achievement of social optimization through the maximization of utility and profit.

The interpretation of environmental degradation, according to Latouche (2013), framed as market failure has driven economic reflection on the effects of the economic process on the natural environment down a false road. This interpretation has shifted to the background of a constituent characteristic of market economies: markets are not the only mechanism for interdependence regulation that is established between individuals through the use of natural capital—and prices are not the only signal that individuals use in their decisions making process. However, there is another element that deserves to be mentioned: any consideration regarding the existence of absolute shortage is discarded.

The utilization of technology, according to Heikkurinen (2018), Likavcan and Scholz-Wackerleb (2016), combined with the mechanisms of the economy, allows an endless process of substitution between environmental goods produced by natural capital and durable goods produced by man (artificial capital). That is the reason why the most optimistic view of neoclassical economists encompasses technological advances as well as the market mechanism, both the price system and economic growth itself, which ultimately would not be subject to any limitations because of the innovative capabilities of the humankind. As said above, Ecological Economy was emerging in the eighties, distancing itself from environmental economy and of course from traditional economy, and positioning itself between ecological science and economics.

The Path of the Ecological Economy

In 1982, Ann-Mari Jansson organized a symposium titled "Integrating Ecology and Economics" in Sweden, in Saltsjobaden. This led Robert Costanza and Herman Daly to decide to start publishing an ad hoc magazine on the topic through a special issue of the Ecological Modelling Journal. (In the introduction to this special issue, Constanza and Daly identify the needs of and the first ecological economy's basic agenda). Due to the success of the two meetings held in Sweden and Poland, a new symposium organized by the well known economist Juan Martinez-Alier was held in Barcelona from 26 to 29 September 1987 entitled "*Integrating Ecology and Economics*".

The meetings produced the idea of starting a specific professional association of ecological economics, and also, the publication of a scientific journal dedicated to in-depth discussion of these issues. In 1988, thanks to the idea of Martinez-Alier, the International Society for Ecological

Economics (ISEE). And in the following year Robert Costanza published the journal "*Ecological Economics*". He was also the first president of the society and first editor of the journal. The first World Conference on Ecological Economics took place in Washington in May 1990, followed by a workshop at the Aspen Institute, attended by 38 ad hoc invited scholars and from which the book, "*Ecological Economics: The Science and Management of Sustainability*", edited by Costanza, arose.

According to Costanza (1991), the EE is a transdisciplinary and interdisciplinary science that sees the natural world as a constantly evolving system, in which the physical laws of thermodynamics and the irreversibility of the transformations of matter and energy operate. It is impossible not to recognize the importance of Georgescu-Roegen and Boulding's theories and the contribution of Herman Daly, pioneer of the discipline well known as the EE. The EE is the bridge that connects Ecology and Economics, such a discipline frames the traditional Economics as an open subsystem of the global ecosystem.

Paraphrasing Costanza (1991), the growing awareness that the system that sustains global ecological life is in danger, forces us to understand that choices made on the basis of local, narrow, and short-term criteria can produce disastrous global results in the long term. We are also starting to realize that the traditional economic and ecological models are not able to fully address global environmental problems. Ecological economy is a new transdisciplinary field that addresses the relationship between the ecosystem and the economic system in the broader sense.

Ecological Economics (EE) according to Ayres and Warr (2010), differs from both economics and conventional ecology by the breadth of its perception of the problems and the importance it attaches to the environment-economy interaction. It is easy to understand how this discipline is transversal and assembles scholars from scientists to economists and chemists, who with their theories have managed to create alternative paradigms. It is of great importance to consider the natural world as an ever-evolving system within which the physical laws of thermodynamics and the irreversibility of material and energy transformations work.

Consequential to this matter is the awareness—emphasized by Cochet (2009), Latouche (2007), Kallis (2011), and Jackson (2011)—of the limit that economic growth is facing as a cause of the negative impact on the natural environment, which is the basis for life on Earth. The ecological economy, according to D'Alisa et al. (2014) raises the issue of a sustainable scale, fair distribution, and efficient allocation. Neoclassical economy deals extensively with allocation, puts in enough considerations regarding distribution but it ignores scale. Scale, according to Kallis et al. (2018), refers to the physical dimension of the economy compared to natural systems. At the passage of the physical volume of "throughput", i.e. the flow, matter-energy from the environment as matter-energy to low entropy and its return to the environment as a waste of high entropy.

If the regenerative capacities of natural systems are jeopardized, continues Kallis et al. (2018), there is no chance of achieving a sustainable flow scale; their regeneration capacity and their assimilative capacities are exceeded, and the waste produced surpasses the metabolization capacity of natural systems. The duty of EE is, according to Kallis et al. (2018), to protect the evolutionary dynamics of ecosystems, maintaining biodiversity and ecological services. Not only that but it protects the fundamental mechanisms of evolution, allows ecosystems to continue the production of their "services" for the survival of our species and keeps alive the extraordinary richness of life on Earth.

Thus, it points out the clear difference between Environmental Economics and Ecological Economics: according to the paradigm of Ecological Economy (Daly, 1977) replacing "natural capital" as part of the "human capital" is not possible, whereas according to Environmental Economics, "human capital" can substitute "natural capital" (weak sustainability). Ecological Economics assumes that "human capital" and "natural capital" are complementary, but not interchangeable. Ecological Economy wants, according to Muraca and Schmelzer (2017), to analyse what limits the Earth's load capacity to the maintenance of the human population. In ecology, carrying capacity is defined as the number of individuals in a population that the resources of the habitat are able to sustain. One of the main goals of EE is, according to Costanza (1991), Muraca and Döring (2018), a solution to the problem of how to measure more fully and exhaustively the present well-being and wealth of our societies. Once the basic notions of microeconomics are going in the right direction, people automatically benefit from it.

The Real Utopia and the Bioeconomics

Currently, for economists to politicians such well-being is measured by the GPD index. GDP is both a sign of market well-being and of well-being of the entire economy (of a country). Perhaps, to summarize in the words of Costanza et al. (1997, p. 89):

> 1. The vision of the earth as a thermodynamically closed and non-materially growing system, with the human economy as a subsystem of the global ecosystem. This implies that there are limits to biophysical throughput of resources from the ecosystem, through the economic subsystem, and back to the ecosystem as wastes; 2. The future vision of a sustainable planet with a high quality of life for all citizens (both humans and other species) within the material constraints imposed by 1; 3. The recognition that in the analysis of complex system like the earth at all space and time scales, fundamental uncertainty is large and irreducible and certain processes are irreversible, requiring a fundamentally precautionary stance; and 4. That institution and management should be proactive rather than reactive and should result in simple, adaptive, and implementable policies based on a sophisticated understanding of the underlying systems which fully acknowledges the underlying uncertainties. This forms the basis of policy implementation which is itself sustainable. (Costanza et al., 1997, p. 89)

1991 was the year of *Beyond the Limits*, a second report produced for the Club of Rome by Donella Meadows, Dennis Meadows and Jorgen Randers. Compared to the first report *Limits to Growth,* the report emphasized that the impact it generated within the community was not strong (one of the reasons being that 20 years is not sufficient time for trends to emerge or to change the thinking on a subject of this kind). According to Turner (2008) the book shows in clear terms the choice we have between a rapid and uncontrolled decline in food production, industrial capacity, population, and life expectancy, or a sustainable future. The authors are respected in the field of systems dynamics for the use of computer models to project the future. They describe a range of possible outcomes and show that a sustainable society is technically and economically feasible.

The conclusions of *Beyond the Limits* while similar to the previous report, are dramatic. Paraphrasing Meadows et al. (1992), humanity had now exceeded the limits of the Earth's ability to sustain its continued existence. The significance of their findings was such that the scholars decided to make it the book's title. They stressed that, at the beginning of the 1990s, overcoming limits could no longer be avoided by enlightened measures: there was no more time. The main objective was to "down" the world, to bring it back to the field of sustainability". The future degrowth paradigm was *in nuce*. However, the issues of degradation and environmental safeguards also began to attract attention around the time, not only in scientific research, but also in the political sphere. Certainly, the pace of the political class is slow compared to pioneers such as K.W. Kapp o N. Georgescu-Roegen, however its contribution, from the point of view of the legislative and environmental framework for the safeguard of the environment was crucial.

Our Common Future: The Political Commitment to Safeguarding the Environment

According to Muraca and Schmelzer (2017), during the early seventies, not only several international organizations of development and cooperation began to include environmental concerns in their agenda, but also national governments (especially the developed countries of the Western part of the world) which created specialized institutions and organizations for the study and prevention of environmental impacts. 1972 was the year of the *United Nations Conference on the Human Environment*, the first international platform which offered the opportunity to discuss environmental issues on a global scale. According to Bardi (2007), this was one of the first attempts of a never-ending journey towards a solution of international environmental problems. The conference clearly demonstrated that there is a battle between national interests and international cooperation and that, at the same time, the existing international order was (and it still is) relatively powerless in solving global issues.

The document set out the following steps: the need for a common outlook and for common principles to inspire and guide the peoples of the world in the preservation and enhancement of the human environment (United Nations, 1973). The preamble of the document outlined the scope of the task ahead: "we must shape our actions throughout the world with a more prudent care for their environmental consequences. Through ignorance or indifference, we can do massive and irreversible harm to the earthly environment on which our life and well-being depend. Conversely, through fuller knowledge and wiser action, we can achieve for ourselves and our posterity a better life in an environment more in keeping with human needs and hopes. [...] To achieve this environmental goal will demand the acceptance of responsibility by citizens and communities and by enterprises and institutions at every level, all sharing equitably in common efforts" (United Nations 1973, cpt. 11).

In addition, the document of the *United Nations Conference on the Human Environment*, contained 26 fundamental principles, all of them related to safeguarding and improvement of natural environment: human liberty and equality; the safeguarding of the natural ecological system, both for present and future generations, through proper planning and administration; restoring, defending, and improving the earth's capacity to produce natural resources; the preservation of wild life and its habitat; the non-exploitation of non-renewable resources; banning the use of non-metabolizable substances; the adoption of measures to prevent pollution from by countries; the allocation and the transfer of financial resources and technical assistance; the adoption of ecological and environmental policies; urban and urban planning; demographic policies; planning, administration, and institutional control over environmental resources; the development and encouragement of scientific and technological research; education; the improvement of international law; international cooperation and, finally, the preservation of the effects of nuclear weapons.

1980 was a cornerstone for ecological awareness thanks to two seminal contributions: *The Global 2000 Report to the President* and *The World Conservation Strategy; Living Resource Conservation for Sustainable Development.* Both studies illustrated alarming situations and provided practical countermeasures to be taken in the shortest possible time, but both also raised a new awareness (especially in the US). The 1980s represented the birth of a new model of development that combined economic growth with equitable distribution of resources. It was the emergence of what would be called sustainable development, strengthened by *The International Union for Conservation of Nature and Natural Resources* who published a document titled *World Conservation Strategy; Living Resource Conservation for Sustainable Development.* This document emphasized the limit that the supply of natural resources presents and the great challenge that humanity will face in the future: the degradation of nature and the environment.

The document was a set of guidelines for environmental issues and on policies that could be implemented both nationally and internationally. 1987 was the year of one of the most important documents related to EE: *Our Common Future* or better known as *The Bruntland Report* (named after the president of the World Commission on Environment and Development who presented it on behalf of the United Nations). The report revolutionized the concept of "sustainable development": "to censure that it meets the needs of the present without compromising the ability of future generations to meet their own needs" (World Commission on Environment and Development, 1987). It was a report which, for the very first time merged the concepts of development, framed as the per capita product growth and the concept of the environment, from which production inputs come and where waste from industrial and consumer activities is returned. This document also brought particular attention to intergenerational dynamics and to the possibility that undisclosed resource exploitation may compromise future well-being. The report also reveals the limits of economic growth, limits which are framed as relative and not absolute, imposed by the state of technology, by social organizations on environmental resources and by the ability of the biosphere to absorb the effects of human activities.

The report includes a section dedicated to "common concerns" regarding the consideration of individual needs and well-being that does not only include economic variables. The second part of the report is dedicated to "collective challenges" in which governments and industries are urged to include instances related to natural resources and the environment in their planning and decision-making processes, so that they can reduce the use of materials and energy; a third part investigated "joint efforts", outlining the hope that environmental protection and sustainable development will become an integral part of the mandates of all government agencies, international organizations, and large private sector institutions. It is possible to interpret the idea of sustainable development as developed by *The Bruntland Report* as an anthropocentric approach aimed at protecting present and future human generations rather than the natural ecosystem. The report raised awareness concerning the interdependencies between economy, ecology, and politics, and the initiatives put in place both by businesses and by governments have multiplied.

Ecological Rationalization: The Urgency to Change Direction

At the same time, however, ambiguities have emerged, including unlimited trust in human technological creativity, institutions and economic-social organization, or the relationship between meeting human needs and protecting the environment. These ambiguities have given rise to the idea, later picked up by Latouche and Hornborg, of "false consent" regarding the mainstream meaning of sustainable development (2009) as an oxymoron. Perhaps its lack of clarity as regards political actions was the reason why it has not been further developed.

As Weart (2003) has argued *The Brundtland Report* does not offer a detailed blueprint for action, but instead a pathway by which the peoples of the world may enlarge their spheres of cooperation. It gives notice that the time has come for a marriage of economy and ecology, so that governments and their people can take responsibility, not just for environmental damage, but for the policies that cause the damage. Sustainable development is not, as Bertorello and Corradi have argued (2011) a fixed state of harmony, but rather a process of change in which the exploitation of resources, the direction of investments, the orientation of technological development, and institutional change are made consistent with future as well as present needs. The commission does not pretend that the process is easy or straightforward. Painful choices have to be made. Thus, in the final analysis, sustainable development must rely on political will.

However, *The Brundtland Report* was the inspirational source of another notable conference: *The Conference of Rio de Janeiro*, organized by the *United Nations Conference on Environment and Development (UNCED)* in 1992. The Conference of Rio closed with one of the most notable documents on environmental protection policies, the *Agenda 21*: a true declaration of principles and a valuable tool for local action. Nelissen et al. (1997, cpt. 2) offer a clear summary of the document: "Every day that goes by without action is a lost one. But how do we design sustainable development? What does a sustainable future mean? What do we need to do to attain that goal? Firstly, we need state regulations".

Those regulations must curb existing unsustainable aspects of society and must shift them in the direction of a sustainable society by means of focused programmes and actions. Furthermore, businesses also need to get involved; they cannot stay out of the debate on the major world issues of environment and development. Businesses must take sides. They must actively work for a lasting future and be compelled to do this in order to stay in business. These actions must not be isolated actions by governments and business, but joint ventures, co-productions. There is a joint responsibility and therefore the need for joint action.

Despite the valid premise, the results achieved by the report, according to Drexhage and Murphy (2010), were quite disappointing. Barkemeyer et al. (2014), emphasized that no surprising decisions were made in Rio de Janeiro. In the phase preceding the conference, there was a lot of lobbying concerning the design texts. Room for new proposals had been diplomatically taken away. The standpoints were known and the parties (G77, G7, US, EC, Japan, etc.) did not give any clue on their own agendas. This new environmental consciousness, forged by *The Bruntland Report,* pushed the European Union to authorize the *Fifth Action Programmeon the Environment*, in order to operationalize the agreements already signed in Rio de Janeiro.

The aim of this programme was to provide a detailed analysis regarding the situation of air, water, soil, waste, quality of life, high-risk activities and biological diversity, and economic features of environmental degradation. The programme also hoped to achieve a change in the behaviour of society, through the participation of all sectors and strengthening of the co-responsibility between public administration, businesses, and the community. The European Union was developing a new strategy for sustainability, focused on agents and economic activities that contribute to the depletion of natural resources and environmental degradation. The action plan specifies the target areas for action: manufacturing, energy, transport, agriculture, and tourism. According to the European Community (1993, pp. 1–4), "the new approach implies, in particular, a reinforcement of the dialogue with industry and the encouragement, in appropriate circumstances, of voluntary agreements and other forms of self-regulation. [...] Energy policy is a key factor in the achievement of sustainable development. [...] The key elements of the strategy up to 2000 will be improvement in energy efficiency and the development of strategic technology programmes moving towards a less carbon-intensive energy structure, in particular, renewable energy options. [...] A strategy for sustainable mobility will require a combination of measures which includes: [...] improved planning, management and use of transport infrastructure and facilities; incorporation of the real costs of both infrastructure and environment in investment policies and decisions, and also in user costs: development of public transport and improvement of its competitive position; [...] promotion of a more environmentally rational use of the private car, including changes in driving rules and habits. [...] to seek to strike a more sustainable balance between agricultural activity, other forms of rural development and the natural resources of the environment. [...] If well planned and managed, tourism, regional development and environment protection can go hand in hand".

The balance between the economic dimension and the ecological dimension was analysed during the late '90s (1997) by the *Kyoto*

Protocol, in which the *United Nations Framework Convention on Climate Change* (UNFCCC) set the achievement to fight global warming by reducing greenhouse gas concentrations in the atmosphere to "*a level that would prevent dangerous anthropogenic interference with the climate system*" (Art. 2). Participants of the UNFCCC signed a Protocol under which OECD countries commit themselves to reducing their emissions of so-called "greenhouse gases" for the period 2008–2012 by 5.2% on average, compared to 1990 levels.

In addition to excluding developing countries from the Protocol's observation, even "emission permits" or "polluting permits" are introduced, a market mechanism that permits trading of emission allowances between a country that has not passed the quota and another that is beyond the established limit. The veto of United States, Canada, Japan and Australia at the Hague Climate Change Conference of 2000 emphasized the difficulties in changing the imagination of the dominant political class.

The Kyoto Protocol entered into force in 2005, despite the non-ratification by the Umbrella Group. Ten years after the first Earth Summit in Rio de Janeiro, The World Summit on Sustainable Development took place in Johannesburg in 2002. This summit recognized the interdependence between environmental sustainability, economic and social components (unlike Kyoto where the two spheres were distinguished). Despite some interesting premises such as globalization management, environmental protection, production, and consumption patterns, the Summit reaffirmed the statements made in Rio de Janeiro in 1992. 2001 saw the European Union again as the main actor of environmental policies (after 1993). At the *VI Piano d'Azione Ambientale* (*Ambiente 2010: il nostro futuro, la nostra scelta*). Four priorities framed by CDC Europee (2001) as the priority of the Union's future environmental policy agenda for the next ten years were identified: climate change, nature and biodiversity, environment and health, sustainable use of natural resources, and waste.

The aforementioned report also underlines the need to make use of renewable and non-renewable resources, and its impact on the environment, is not over-emphasized by its ability to load, to dissociate the use of resources from economic growth and to improve it resource efficiency through dematerialization of the economy and prevention of waste generation.

The "Happy" Degrowth

No literature review which looks at happy degrowth, bien vivir, sustainability, and so on, is complete without mentioning one of the pioneers of this change of direction in the collective imagination due to his principled approach in relation to current political, social, and economic thought: Serge Latouche.

Basing his conceptual approach on the foundations laid by Boulding and other economists who started the EE era, the French economist argues that a system-oriented towards continuous growth of production and consumption does not produce an effective increase in social welfare. Latouche sees mankind as threatened by the concept of ever-increasing production and consumption which causes both environmental as well a social degradation. A pillar that sustains the theoretical basis of degrowth is an anthropological approach, that very often betrays Latouche's education in Economics.

The fusion of the anthropological and economic sphere has brought Latouche to develop the concept of "Happy Degrowth", i.e. a kaleidoscope of alternatives (some feasible, others utopian), founded on human values and their intertwining with structures and economic and social institutions, trying to provide tools that can be useful for a transition leading to degrowth: an alternative, desirable society that understands the limits of our planet and in order to protect nature, protects also mankind by observing a clear distinction between well-having and well-being. As we will see in the development of this chapter, it is the lack of balance between the economic and anthropological approaches that makes the work of the French professor subject to criticism and factual limits.

"We do not claim to have a recipe for the future…but we can no longer pretend that we can keep growing as if nothing has happened… Is no longer possible having an infinite growth in a finite planet…" These three statements synthesise the Second International Conference on Economic Degrowth for Ecological Sustainability and Social Equity, which took place in Barcelona in 2010. However, as previously outlined in this chapter, it was Georgescu-Roegen in 1972 during a conference titled Energy and Economic Myths, who conceptualized the first embryonic development of what was to become degrowth theory.

Georgescu-Roegen's, analysis of the available literature from Pigou to the stationary state of Herman Daly, including his critical evaluation of Mill's vision undertaken in order to find a solution to the impossibility of infinite growth on a finite planet, came to the conclusion that "the necessary conclusion of the argument in favour of that vision is that the most desirable state is not a stationary, but a declining one. Undoubtedly, the current growth must cease, nay, be revisited. But anyone believes that he can draw a blueprint for the ecological salvation of the human species does not understand the nature of evolution, or even of history-which is that of a permanent struggle in continuously novel forms, not that of a predictable, controllable physico-chemical process, such as boiling an egg or launching a rocket to the moon" (Georgescu-Roegen, 1971, p. 369).

As was demonstrated, Georgescu-Roegen initially used the word "declining"; it was Jacques Grinevald 1995 translation of the Romanian economist's works that coined the term theory of "degrowth". At the end of the 1990s Vincent Cheynet and Bruno Clémentin established the association Casseurs de Pub came to life. This association proposed a major challenge. To stimulate creativity and to alert public opinion to the dangers of capitalism and its commodities, the goal of the Casseurs de Pub was to find a possible way of thinking about sustainability. It was during 2000 that the interest for this new theory reached significant levels.

In 2002, the UNESCO in Paris was the stage for the international symposium organized by Serge Latouche and by La Ligne d'horizon: Défaire le dévelopment, refaire le monde. However, that was not all. In the same year the scientific, ecological journal Silence dedicated a special issue to degrowth theory based on teamwork by Latouche, Clémentin and Cheynet. The convention was an outstanding achievement which led to a second international congress in Lione in 2003. The Décroissance soutenable was the product of the joint effort between Silence, L'Ecologiste, and Casseurs de Pub. This was also the occasion of the launch of the book *Objectif Décroissance*. The following year, 2004, saw the publication of La Décroissance, the first journal dedicated to this new discipline. The success of the journal was immediate and went beyond the expectations of the publisher and intellectuals involved.

What is degrowth? According to Latouche (2007), Martínez-Alier (2009), Kerschner, (2010), Weiss and Cattaneo (2017), Degrowth is the deconstruction of political economy through an economic-anthropology approach. It is to question the *homo economicus*, and Latouche retraces the steps of K. Polanyi, M. Sahlins, H. D Thoreau, and M. Mauss. In *Critique de l'Imperialisme* (1979), Latouche argues that the social reality of an economic anthropology object was very alien to economists, and the challenge was to question them, through an interpretation of development and underdevelopment as "deculturation" as per Leninist theory of imperialism.

The Occidenatalisation du Monde and the Faut-il Refuser le Dévelopement? are not contaminated yet by the ecological sphere but follow in the footsteps of the first book. It was due to the La Planète des Naufragés that we can frame Latouche as close to the position of the Club of Rome. Latouche finds his dimension in which was acknowledged such as a little international masonry led by Ivan Illich, in which gathered students and followers such as Latouche and Sachs. As Barbara Muraca and Schmelzer argues (2017), the framework behind the idea of development at that time was always related to the south of the world, because it was the north that developed the south. The turning point in Latouche's work, L'Autre Afrique: Entre don et marché (1998), speaks of the great excluded of the world and their organization trying to survive.

In this book, inspired by the work of Zerzan, Salhins, and Mauss, Latouche focuses on systems in Laos, and explains how these people survived outside the economy due to creativity and self-organization. According to Latouche (1998), this model should provide us with guidelines on what could be a different idea of growth or a way to abandon it, living with less material goods and but more goods capable of bringing joy of life. Degrowth is not a concept or a complex theory borrowed from the economics of growth. For some scholars, Degrowth is an idea, and we can find its core in the broad area of political sciences, social studies, philosophy, ecology, and so on. On the other hand, it can be argued that we can label Degrowth as a political movement, captured in a provocative slogan.

Paraphrasing Latouche (1979), pioneer of the Décroissance, the term degrowth contains an older history, linked to critical "culturalist" economy, and on the other hand, its environmentalist critics. According to the French economist, the merger of these two types of criticism has created the foundation for Degrowth. Degrowth, like all matters that create a disturbance in our collective imaginary, based on the ideologies of capitalism and ultraliberalism, is perceived with fear and mystifies the financial wizards and by the world's leaders who have always followed the slogan of TINA (there is no alternative). A society based on growth follows simple rules, quasi simple as an Aristotelian syllogism.

The Convenient Lie

Our economic system is built upon growth; economic growth is the only path of mainstream macroeconomics; all economic policies are designed to achieve growth with a frantic pace. Again, there is no alternative. For the standard macroeconomics, a "no growth economy" is an ungrowth, a de-development, an oxymoron. However, degrowth is not recession, it is not a GDP destroyer (fetish indicator of our living conditions and well-being), Degrowth is a redesign of our collective imagination that implies a priori the complete transformation of our economic system. Degrowth is a defiance that goes beyond the physical dimension of the economic process.

It is very important to stress that this theory has its basis in the critique of the "gross domestic product" as an imperfect measure of well-being and the common view that welfare is measurable through the consumption and quantity of acquiring goods. According to Fotopulos (1999), who understands the lesson of Joseph Schumacher, when a society's only aim is producing more and more without considering the impact on nature, the economic system is oriented, at any level, to the maximization of economic growth. The 2000s were very productive in terms of research on Degrowth (Daly, 2010; Jackson, 2011; Kallis, 2011; Kallis et al., 2009; Kerschner, 2010; Martinez-Alier, 2009; van den Bergh, 2011; Kallis et al., 2012; Victor, 2014).

These different writers share the positive idea that Degrowth as an opportunity to replace material abundance by a better quality of life with stronger social, local and natural relationships. According to Jackson (2011) and Latouche (2007) and other scholars, an undefined pursuit of growth is not compatible with a finite planet. It is imperative to change direction, a turnaround that involves the abandonment of the goal of unlimited growth, driven exclusively by the pursuit of profit of the capital holders. They also share a large, holistic, systemic approach that can be considered a macroeconomic Degrowth project (Martínez-Alier et al., 2010). As predicted by the Club of Rome, the creation of a more coherent and desirable society, based on Degrowth, appears not only as a real possible exit from consumer society, but also a political goal in the short term.

If it is true that there is no worst-case scenario of a society based on growth without growth, the workable alternative will be consuming less and more consciously to live better. Referring to the thoughts of authors such as Karl Polanyi, Marshall Sahlins and Marcel Mauss, Degrowth theory could be synthesized in the metaphor of the snail of Ivan Illich: the snail not only teaches us the virtue of slowness but provides another, even more important lesson. The snail constructs the delicate architecture of its shell by adding ever-increasing spirals one after the other, but then it abruptly stops and winds back in the reverse direction. In fact, just one additional larger spiral would make the shell sixteen times bigger. Instead of being beneficial, it would overload the snail.

Any increase in the snail's productivity would only be used to offset the difficulties created by the enlargement of the shell beyond its preordained limits. Once the limit to increase spiral size has been reached, the problems of excessive growth multiply exponentially, while the snail's biological capability, in the best of cases, can only show linear growth and increase arithmetically. The ability of the snail to abandon the exponential growth it adopted for a certain time shows how we can imagine a Degrowth society, calm and convivial and respectful of the trinity of traditional knowledge, culture and cuisine.

One of the main criticisms of Degrowth comes from Van den Bergh (2011), who has emphasized that Degrowth (interpreted as a radical process that involves the degrowth of consumption, worktime, and so on) is environmentally ineffective, "socially and politically unfeasible, and economically inefficient". Van den Bergh (2011, p. 881) has stressed that these multiple interpretations of Degrowth make it an "ambiguous and rather confusing concept". For Kallis (2011) this is an oversimplification, but perhaps, the answer is the foundation of a different collective imaginary that abolishes and overcomes the founding categories of the economy. The crisis we are experiencing nowadays is a society of growth without growth, is a crisis of civilization, is a social and civilian decline and, according to some authors, starting with Latouche, it will inevitably lead to barbarism.

Contemporary economic, political, social, and environmental crises are an inevitable result, servitude to the lobbies of multinational marketing and international finance is emptying the contents of our democratic society. According to Gare (2002, p. 116), "The consequence of the internalization of the market economy and the concentration of economic power it engenders is an ecological crisis that threatens to develop into an eco-catastrophe, the destruction of the countryside, the creation of monstrous mega-cities and the uprooting of local communities and cultures". Crouch posits that we live in a "post-democracy" (2008) in which political consciousness and responsibility of citizens are obscured and marginalized by the impossibility to choose our representatives, by the failure to recognize the outcome of referenda, by unidirectional strictness of "technical governments", by partisan factions which continue to defend strenuously their consolidated privileges.

According to Johan Galtung (1969), the characteristic features of Western cosmology are the following: Western conception of space, centralized, universal; linear time focused on the present; analytical and not holistic approach of epistemology; conception of human relations in terms of domination. The imaginary foundations of the institution of the economy must be called into question. Newfound frugality allows the reconstruction of a society based on what Ivan Illich called "modern subsistence". It seems possible to try to summarize the meaning, the ultimate achievement of a virtuous circle—synthesized by Latouche in his 8 Rs-with the words of Cornelius Castoriadis:

> it is necessary that profound changes take place in the psychosocial organization of Western man, with respect to his behaviour towards life, in his imagination. It is necessary that the idea that the only purpose of life is to produce and consume—an idea both absurd and degrading—is abandoned; it is necessary that the capitalist imagination of a pseudo-rational pseudo-control, of an unlimited expansion, be abandoned. Only men and women can achieve all this. A single man or organization can at most prepare, criticize, incite, possible orientations. (Castoriadis, 2005, p. 244)

The degrowth path will not easily erase either e-money, or markets or wage labour. For instance, Plato, in his "The Republic", sees the importance of the exchange as part of the functioning of republican institutions. To the question, "How will be made the exchange of goods that everyone produces?" The philosopher answered that this is the reason why humankind has founded a community and a state; the exchange will take place through sales and purchases—we will have as a result, a market and a currency, conventional symbol to make an exchange (Dacrema, 2010). On the other hand, for Aristotle, wisdom is an engine that triggers the virtuous circle: exchange and a monetary system based on the economy are necessary for the community (koinonia), but only if such tools are used within certain limits. Perhaps Aristotle has offered the first framework of the Buen Vivir. For the philosopher, the sense of limitation is an imperative. It is necessary to prevent the implosion of our economic system but in addition, it is a social thermometer to ensure social justice in order to achieve the *eudaimonia*.

Through an observation of the ideologies of Aristotle, crisis, triggering element of degrowth/resilience, is the capacity to discern (*krino*: i.e. crisis) between a good or bad utilization of goods. The exchange thus, is a necessary tool for completing the natural self-sufficiency of humankind. When the exchange degenerates turning it into an uncontrolled accumulation of wealth the economy is emptied of its ethical and social principles. For Aristotle it is not possible to achieve the Buen Vivir, through the uncontrolled accumulation of goods; the bonds of solidarity and reciprocity are broken within the community: natural economy becomes economics against nature. The transition to what we call "*DemocraCity*" will involve new and hybrid social and economic measures with the purpose of taking care of the community, creating more durable goods, improving the stages of repair and recycling.

The boost in employment is no longer sustainable through a consumption society. It is fundamental to achieve, according to Latouche (2004), in order to exit from the capitalist system of growth, a drastic reduction in hours worked per week in order to ensure that everyone has a satisfactory job. At the same time, this measure could achieve the necessary reduction of the consumption of natural resources. It seems definitely possible that a degrowth programme at the beginning thus, can lead to an increase of production due to the demand for products and environmentally friendly machinery. An economy based on renewable energy would mean the development of productive sectors such as the construction of windmills, turbines, and related spare parts; furthermore, it would also entail the construction of hydrogen engines, bicycles and light rail in addition to a widespread increase in organic farming and reforestation. The reduction, reuse, repair, and recycling process that could be achieved by abandoning planned obsolescence, could increase employment and boost the creation of new and resilient jobs Latouche argues, (2004) such as the "Eco Architect" or the "Silvicultural Expert".

The Circle of the 8Rs

Starting from the imperative of responsibility raised by Hans Jonas (1985), Umberto Eco (2004) and Tomunubo Imanichi (2009), Serge Latouche traces the ideological programme of his utopia with the famous 8Rs (Re-evaluate, Re-conceptualize, Restructure, Redistribute,

Relocalize, Reduce, Re-use, and Recycle). These require explanation and definitions:

1. Re-evaluate: revive the importance of cooperation, the pleasure of leisure and play, the importance of social life, of autonomy, of the taste for beauty and of the reasonable. The recovery of these values allows us finally to move from a predatory attitude of domination over nature to a harmonious integration into the natural world.
2. Re-conceptualize redefine the concepts of wealth and poverty and deconstruct the concepts of rarity and abundance. It creates an artificial need, commercializing nature.
3. Restructure: abandon the capitalist system.
4. Redistribute: redistribution is configured as a new mode of wealth distribution and access to the natural heritage between the North and South of the world and, within each society, between classes, individuals and generations.
5. Relocalize: acquire products that we need to satisfy the needs of local populations, limiting the movement of goods and capital. Even politics, culture, lifestyle have to be rethought in this way. Every decision, political, economic and so on, will be taken locally.
6. Reduce: limit overconsumption and the huge waste created by our lifestyle.
7. (includes 8) Re-use and Recycle: two possible and desirable ways to reduce our uncontrolled waste.

The virtuous circle of the 8Rs contains simultaneously innovations and repetitions. They are revolutionary and historical acts which indicate possible workable solutions. A cornerstone of this project is achieving autonomy as in its etymological meaning, autónomos i.e. independent, in contrast to the heteronomy of the "invisible hand" of the market, of the dictatorship of financial markets and in contrast to the "techno-knowledge" of the "ultra" modern society. The systematic use of the small fragment "re" in the 8Rs is not a romantic thought or a return to the past, or even a conservative thought. According to Latouche (2008), the 8Rs are the concrete answers towards a development system, such as the capitalist system that utilises many "over" concepts including overdevelopment, overabundance, overconsumption, and overindebtedness.

References

Adkisson R. V. (2009). The economy as an open system: An institutionalist framework for economic development. In T. Natarajan, W. Elsner, & S. Fullwiler (Eds.), *Institutional analysis and praxis*. New York, NY: Springer. https://doi.org/10.1007/978-0-387-88741-8_2

Andrew, B. (2008). Market failure, government failure and externalities in climate change mitigation: The case for a carbon tax. *Special Issue: Symposium on Climate Change, Governance and Environmental Services, 28*(5), 393–401.

Anguelovski, I. (2015). Tactical developments for achieving just and sustainable neighborhoods: the role of community-based coalitions and bottom-to-bottom networks in street, technical, and funder activism. *Environment and Planning C: Government and Policy, 33*(4), 703–725. https://doi.org/10.1068/c12347

Arrow, K., et al. (1995). Economic growth, carrying capacity, and the environment. *Ecological Economics, 15*(2), 91–95.

Ayres, R., & Warr, B. (2010) *The economic growth engine: How energy and work drive material prosperity (The International Institute for Applied Systems Analysis)*. Edward Elgar Publishing.

Bardi, U. (2007). Energy prices and resource depletion: Lessons from the case of whaling in the nineteenth century. *Energy Sources*, Part B, *2*(3), 297–304.

Barkemeyer, R., Holt, D., Preuss, L., & Tsang, S. (2014). What happened to the 'Development' in sustainable development? Business Guidelines Two Decades After Brundtland. *Sustainable Development, 22*(1), 15–32.

Becker, C. (2006). The human actor in ecological economics: Philosophical approach and research perspectives. *Ecological Economics., 60*, 17–23.

Berger, S. (2008). K. William Kapp's theory of social costs and environmental policy: Towards political ecological economics. *Ecological Economics*, *67*(2), 244–252.

Bertorello, M., & Corradi, D. (2011). *Capitalismo tossico. Crisi della competizione e modelli alternativi*. Edizioni Alegre.

Bianchi, J. (2011). Overborrowing and systemic externalities in the business cycle. *American Economic Review, 1*(7), 3400–3426.

Bogdanov, A. (1921). *Očerki Vseobščej Organizacionnoj Nauki* [Essays in the Universal Organizational Science]. Samara.

Boulding, K. (1966). The economics of knowledge and the knowledge of economics. *American Economic Review, 56*(1/2), 1–13.

Brougher, K. (2015). Art and nuclear culture. *Bulletin of the Atomic Scientists, 69*(6), 11–18.

Burke, P. J., Shahiduzzaman, M., & Stern, D. I. (2015). Carbon dioxide emissions in the short run: The rate and sources of economic growth matter. *Global Environmental Change, 33*, 109–121.

Cacciari, P. (2006). *Pensare la decrescita. Sostenibilità ed equità*. Intra Moenia.

Calafati, A. G. (1997). Il degrado ambientale come fallimento del mercato? *La questione agraria*, 67.

Carson, R. (1962). *Silent spring*. Houghton Mifflin.

Castoriadis, C. (2005). *Une sociétè à la dérive* (p. 244). Seuil.

Cochet, Y. (2009). *Antimanuel d'écologie*. Breàl – Parution.

Commoner, B. (1986). *Il cerchio da chiudere. Presentazione di Giorgio Nebbia Appendice di Virginio Bettini*. Garzanti Libri.

Conti, L. (1987). *Influenze e tendenze nello sviluppo della cultura ecologica*. Russo e Silvestrini.

Costanza, R. (1991). *The science and management of sustainability: Ecological economics*, Columbia University Press, New York.

Costanza, R., Cumberland, J., Daly, H., Goodland, R., & Norgaard, R. (1997). An introduction to Ecological Economics. *International Society for Ecological Economics*, Boca St. Lucie Press, Raton (Florida).

Crouch, C. (2000). *Coping with post democracy*. Fabian Pamphlets.

Crouch, C. (2008). What Will Follow the Demise of Privatised Keynesianism? *The Political Quarterly, 79*(4), 476–487.

Dahlman, C. J. (1979). The problem of externality. *The Journal of Law and Economics, 22*(1), 141–162.

D'Alisa, G., Demaria, F., & Kallis, G. (2014). *Degrowth: A vocabulary for a new era*. Routledge.

Dacrema, P. (2010). *Il miracolo dei soldi. Come nascono, dove vanno, come si moltiplicano*. Etas.

Daily, G., Postel, S., Bawa, K. S., & Kaufman, L. (1997). *Nature's services: Societal dependence on natural ecosystems*. https://www.researchgate.net/publication/37717461_Nature's_Services_Societal_Dependence_On_Natural_Ecosystems

Daly, E. H. (1977). *Steady-state economics*, Island Press, Washington.

Daly, H. (1981). *Lo stato stazionario. L'economia dell'equilibrio biofisico e della crescita morale, Sansoni*. Firenze.

Daly, H. E. (1992). Allocation, distribution, and scale: towards an economics that is efficient, just, and sustainable. *Ecological Economics, 6*(3), 185–193.

Deese, R. S. (2009). The artefact of nature: 'Spaceship Earth' and the dawn of global environmentalism. *Endeavour, 33*(2), 70–75.

Drexhage, J., & Murphy, D. (2010). *Sustainable development: From Brundtland to Rio 2012*. United Nations Headquarters, 1st Meeting by the High-Level Panel on Global Sustainability, United Nations, New York.

Eco, U. (2004). *Storia Della Bellezza*. Rizzoli.

Ehrlich, P., & Ehrlich, A. (1969). *The population bomb*. Ballantine.

Esposito, M., Tse, T. & Soufani, F. (2018). The circular economy: An opportunity for renewal, growth, and stability. *Thunderbird International Review, 60*(5), 725–728.

European Community. (1993). *Fifth action programme on the environment.* Bruxelles.
Fisher, W. F., & Ponniah, T. (2003). *Another world is possible: Popular alternatives to globalization at the world social forum.* Zed Books.
Fotopulos, T. (1999). *Per una Democrazia Globale.* Eleuthera, Milano.
Galtung, J. (1969). Violence, peace, and peace research. *Journal of Peace Research, 6*(3), 167–191.
Gare, A. (2000). Aleksandr Bogdanov's history, sociology and philosophy of science. *Studies in History and Philosophy of Science, Part A, 31*(2), 231–248.
Gare, A. (2002). Towards an inclusive democracy: The crisis of the Growth Economy and the Need for a New Liberatory Project, by Takis Fotopoulos. *Review of Radical Political Economics, 34*(1) (34), 97–99.
Georgescu-Roegen, N. (1971). *The entropy law and the economic process.* Harvard University Press.
Glasson, J. (2012). *Introduction to environmental impact assessment.* Routledge.
Goldsmith, E. (1972). A blueprint for survival. *The Ecologist*, n. 2.
Golub, R., & Townsend, J. (1977). Malthus, multinationals and the club of Rome. *Social Studies of Science, 7*(2), 201–222.
Graeber, D. (2011). *Debt: The first 5,000 years.* Melville House.
Hardin, G. (1968). The tragedy of the commons. *Science, 162,* 1243–1248
Heidenreich, R. (1998). Economics and institutions: The socioeconomic approach of K. William Kapp. *Journal of Economic Issues, 32*(4), 965–984.
Heidenreich, M. (2016). *Exploring inequality in Europe. Diverging income and employment opportunities in the crisis.* Edward Elgar Publishing Limited.
Heikkurinen, P. (2018). Degrowth by means of technology? A treatise for an ethos of releasement. Part 2, *Journal of Cleaner Production, 197*, 1654–1665, Elsevier.
Helmreich, S. (2011). From spaceship earth to Google Ocean: Planetary icons, indexes, and infrastructures. *Social Research, 78*(4), 1211–1242.
Helmsing, B. (2001). Externalities, learning and governance: New perspectives on local economic development. *Development and Change, 32*(2), 277–308.
Höhler, S. (2007). Carrying capacity—The moral economy of the coming spaceship earth. *Atenea: A Bilingual Journal of the Humanities and Social Sciences XXVI,* 1, 59–74.
Imanichi, T. (2009). *An introduction to eco-ethica.* UPA.
Jackson, T. (2011). *Prosperity without growth: Economics for a finite planet.* Earthscan, London.
Jackson, T., & Webster, R. (2016). Limits revisited. *A review of the limits to growth debate.* Creative Commons.
Jaffe, A. B., Newell, R. G., & Stavins, R. N. (2005). A tale of two market failures: Technology and environmental policy. *Ecological Economics, 54*(2–3), 164–174, Elsevier.

Jonas, H. (1985). *The imperative of responsibility: In search of an ethics for the technological age*. University of Chicago Press.
Jones, P. (2010). Responding to the ecological crisis: Transformative pathways for social work education. *Journal of Social Work Education, 46*(1), 67–84.
Kallis, G. (2011). In defence of degrowth. *Ecological Economics, 70*, 873–880.
Kallis, G. (2017). Economics without growth. In M. Castells (Ed.), *Another economy is possible: Culture and economy in a time of crisis* (pp. 34–54), Cambridge University Press.
Kallis, G., Kerschner, C., & Martinez-Alier, J. (2009). The economics of degrowth. *Ecological Economics, 84*, 172–180.
Kallis, G., Kerschner, C., & Martinez-Alier, J. (2012). The economics of degrowth. *Ecological Economics*. https://degrowth.org/wp-content/uploads/2012/11/Kallis_2012_The-economics-of-degrowth.pdf
Kallis, G., Kostakis, V., Lange, S., Muraca, B., Paulson, S., & Schmelzer, M. (2018). Research on degrowth. *Annual Review of Environment and Resources*.
Kallis, G. Martinez-Alier, J., & Schneider, F. (2010). Growth, recession or degrowth for sustainability and equity? *Journal of Cleaner Production*.
Kapp, K. W. (1971). L'economia come sistema aperto. *Economia e ambiente, Saggi scelti*. Otium Edizioni, Ancona.
Kempf, H. (2010). *Per salvare il pianeta dobbiamo farla finita con il capitalismo*. Garzanti.
Kerschner, C. (2010). Economic de-growth vs. steady-state economy. *Journal of Cleaner Production, 18*, 544–551.
LaRouche, L. (1983). *There are no limits to growth*. New Benjamin Franklin House.
Latouche, S. (1979). *Critique de l'impérialisme. Une analyse marxiste non léniniste de l'Impérialisme*. Anthropos, Paris.
Latouche, S. (1998). *L'Autre Afrique: Entre don et marché*. Albin Michel.
Latouche, S. (2004). Degrowth economics. *Le monde diplomatique*. https://www.jussemper.org/Resources/Economic%20Data/Resources/Degrowth%20economics,%20by%20Serge%20Latouche.pdf
Latouche, S. (2006). *The globe downshifted*. Le monde diplomatique.
Latouche, S. (2007). *La scommessa della decrescita*. Feltrinelli.
Latouche, S. (2009). *Farewell to growth*. Polity Press.
Latouche, S. (2013). Prefazione in Bonaiuti, M *La grande transizione: Dal declino alla società della decrescita*. Bollati Boringhieri, Torino.
Likavcan, L., & Scholz-Wackerleb, M. (2016). Technology appropriation in a de-growing economy. *Journal of Cleaner Production, 197*, 1666–1675.
Martínez-Alier, J. (2009). Socially sustainable economic de-growth. *Development and Change, 6*(40), 1099–1119.

Martínez-Alier, J., Pascual, U., Vivien, F-D., & Zaccai, E. (2010). Sustainable degrowth: Mapping the context, criticisms and future prospects of an emergent paradigm. *Ecological Economics, 69*(9), 1741–1747.

Mattauch, L., Hepburn, C., & Stern, N. (2018). *Pigou pushes preferences: Decarbonisation and endogenous values* (CESifo Working Paper no. 7404).

Matthey, A. (2010). Imitation with intention and memory: An experiment. *Journal of Behavioral and Experimental Economics* (formerly *The Journal of Socio-Economics*), *39*(5), 585–594.

Meadows, D, et al. (1992). *Beyond the limits: Global collapse or a sustainable future*. Earthscan Publications.

Minear, R. H. (2011). *The day the Sun rose in the west: Bikini, the Lucky Dragon, and I*. Hawaii.

Muraca, B., & Döring, R. (2018). *From (strong) sustainability to degrowth: A philosophical and historical reconstruction*.

Muraca, B., & Schmelzer, M. (2017). Sustainable degrowth: Historical roots of the search for alternatives to growth in three regions. In I. Borowy & M. Schmelzer (Eds.), *History of the future of economic growth: Historical roots of current debates on sustainable degrowth* (pp. 174–198), Routledge, London.

Nelissen, N., Van Der Straaten, J., & Klinkers, L. (1997). *Classics in environmental studies. An overview of classic texts in environmental studies*. International Books, Utrecht.

Odum, E. P. (1969). The strategy of ecosystem development. An understanfing of ecological succession provides a basis for resolving man's conflict with nature. *Science, 164*, 262–270.

Odum, E. P. (1971). *Fundamentals of ecology* (3rd ed.), 1–574, W.B. Saunders Co, Philadelphia.

Pigou, A. C. (1920). *The economics of welfare*. Macmillan.

Piketty, T. (2013). *Capital in the twenty-first century*. Harvard University Press.

Pimentel, D., Wilson, C., McCullum-Gomez, C., & Huang, R. (1997). Economic and environmental benefits of biodiversity. *BioScience, 47*(11), 45–55.

Redclift, M. (2002). *Sustainable development: Exploring the contradictions*. Routledge.

Reid, L. (1962). *The sociology of nature*. Penguin Books.

Sen, A. (2011). *The Idea of Justice*. Harvard University Press.

Siebenhüner, B. (2000). Homo sustinens—Towards a new conception of humans for the science of sustainability. *Ecological Economics, 32*, 15–25.

Solow, R. (1974). Intergenerational equity and exhaustible resources. *Review of Economic Studies, Symposium on the Economics of Exhaustible Resource, 41*, 29–45.

Stiglitz, J. E. (2010). *Risk and global economic architecture: Why full financial integration may be undesirable*. Nber Working Paper Series.

Swaney, J. A., & Evers, M. A, (1989). The social cost concepts of K. William Kapp and Karl Polanyi. *Journal of Economic Issues, 33*(1), 7–**33**.
United Nations. (1973). *Report of the conference on the human environment* Stockholm, 5–16 June 1972, New York.
Turner, G. M. (2008). A comparison of the limits to growth with 30 years of reality. *Global Environmental Change, 18*(3), 397–411.
Valentinov, V., & Chatalova, L. (2013). Transaction costs, social costs and open systems: Some common threads. *System Research and Behavioral Sciences, 31*(2), 316–326.
Vatn, A. (1998). Input versus emission taxes: Environmental taxes in a mass balance and transaction costs perspective. *Land Economics, 74*(4), 514–525.
van den Bergh, J. C. J. M. (2011). Environment versus growth—A criticism of "degrowth" and a plea for "a-growth". *Ecological Economics, 70*(5), 881–890, Science Direct, Elsevier.
Victor, P. A. (2014). The Kenneth E. Boulding memorial award 2014: Ecological economics: A personal journey. *Ecological Economics, 109*, 93–100, Science Direct, Elsevier.
Weart, S. R. (2003). *The discovery of global warming*. Harvard University Press.
Weiss, M., & Cattaneo, C. (2017). Degrowth – Taking stock and reviewing an emerging academic paradigm. *Ecological Economics, 137*, 220–230.
World Commission on Environment and Development. (1987). *Our common future*. Oxford University Press.

CHAPTER 3

Providing a Concrete Alternative: The Transition Towards the Post-development

The goal of this chapter is to provide an historical overview of the background starting with analysis by Serge Latouche and other facets of growth. The Degrowth paradigm is often surrounded by a methodological confusion. The project to exit the consumer society in order to build a better society, called "frugal abundance" by Latouche (2012, p. 21) cannot fail to create misunderstandings, raise objections and resistance, whatever the course and path Degrowth takes. Abandoning the economy of the infinite growth and changing the collective imagination is what the project of Degrowth is attempting to achieve.

What is Degrowth then? Degrowth is deconstruction of the current political economy through an economic anthropology approach. It is to question the *homo oeconomicus*, and here Latouche retraces the steps of K. Polanyi, M. Sahlins, H. D. Thoreau and M. Mauss. In the *Critique de l'Imperialisme* Latouche argued (1979) that the social reality of the economic anthropology object was totally alien to economists and the challenge was to question them, through an interpretation of development and underdevelopment as "deculturation" of the Leninist theory of imperialism. Degrowth, according to Aries (2006), is not an economic theory that wants to reverse the crisis, but a life practice that uses the crisis to dismantle and rebuild society itself.

This is the reason why Degrowth theory differs from other theories proposed to resolve the stalemate in which the growth society operates

M. Carcea, *The Road Towards DemocraCity*,
https://doi.org/10.1007/978-981-16-3505-2_3

and with which it must not be confused. Latouche is the one, more than other scholars, that has contributed to the construction, development of the "provocation", i.e. Degrowth theory, which stresses the urgency of an inversion and a radical change in the contemporary socio-economic imagination. The French economist began his paradigm from an empirical assumption that a system based on an infinite growth on a finite planet, a system based on demonical circle of growth, production, and consumption, does not produce an effective improvement in the standard of living.

The future of mankind itself is in jeopardy by a logic directed to produce and consume more than is really needed and which causes both environmental and social degradation. The anthropological dimension is a cornerstone of the author's ideological development, more so than his education as an economist. The fusion of the two spheres, the anthropological and economical one, is one of the pillars of Latouche's paradigm. His alternative proposal is the result of a mixture of themes, such as concepts borrowed from economics and applied to human values.

As the relevant literature demonstrates, the anthropological dimension of Degrowth represents a weakness from an economic point of view. However, Latouche's work constitutes the reference point and the fundamental conceptual basis for understanding and embarking on the path that will bring us to *DemocraCity*. However, the road that will lead us towards this city of concrete utopia must start from the analysis of the realities of the Transition Town. The movement of Transition Towns is perhaps the form of construction from below, that most depicts the nature of a Degrowth society. These cities which have embarked on Transition Town trajectories, are looking for energy self-sufficiency from the perspective of having awareness of the finite nature of fossil fuels. More generally they seek the application of and an increase in resilience. This concept, borrowed from physics and encompassing ecology, can be defined as the ability of an ecosystem to resist sudden and violent changes.

For example, how can large urban areas cope with the end of oil, the rise in temperature, and the attendant catastrophes as predicted by Hubbert (1949). The answer from the point of ecological experience, according to Latouche (2011), is that while specialization allows the improvement of performances in the field, it decreases the resilience of the whole. Diversity though, Latouche argues (2011), reinforces resistance and the ability to adapt. Reintroducing vegetables, polyculture, proximity agriculture, small craft units, multiplying the sources of renewable energy, all these activities together reinforce resilience.

Farewell Global Warming, Welcome to Climate Change!

According to The United Nations Framework Convention on Climate Change UNFCCC (1992), the term climate change refers to a statistically significant variation both of the average temperatures of the climate variability, persisting for an extended period. Climate change is therefore a modification of the average temperature that characterizes a given region. When we talk about global climate change, we are referring to changes in the climate of Earth as a whole, including phenomena such as rising temperatures (global warming) or decreasing temperatures, rather than permanent changes to winds and currents. Scientifically, scholars and institutions (Karl & Trenberth, 2003; Bardi, 2007; Bardi & Lavacchi, 2009; Ali, 2013; Hertzberg, 2017) strongly agree that climate change stems from anthropogenic causes and that greenhouse gas emissions will continue to rise as long as human lifestyle is dependent on oil.

The Intergovernmental Panel on Climate Change (IPCC) Fifth Assessment Report (2014) stressed that the facts about climate warming are unequivocal and, since the 1950s, many of the changes observed are unprecedented on time scales that vary from decades to millennia. The atmosphere and the oceans have warmed up, snow, and ice have been reduced, sea levels have risen, and the concentrations of greenhouse gases increased. According to the report, the surface atmospheric temperature shows that each of the last three decades on the Earths surface has been sequentially warmer than any previous decade since 1850. In the northern hemisphere, according to Van Ypersele (2015), the 1983–2012 period was probably the hottest thirty years of the last 1400 years.

Over the last twenty years, according to the IPCC Fifth Assessment Report, the ice caps of Greenland and Antarctica have lost their mass, the glaciers have continued to retreat on almost the whole planet, while the extension of the Arctic sea ice and the spring snow cover in the northern hemisphere have continued to decline in extension. The rate of sea-level rise since the mid-nineteenth century has been larger than the average rate of the previous 2000 years. In the period 1901–2010, the average global sea level grew by 0.19 (0.17–0.21). Atmospheric concentrations of carbon dioxide, methane, and nitrous oxide have increased to unprecedented levels at least compared to the last 800,000 years.

The concentration of carbon dioxide has increased by 40% from the pre-industrial age, primarily for emissions related to the use of fossil fuels, and in the second instance for the net emissions linked to the change in land use. Oceans have absorbed about 30% of the anthropogenic carbon dioxide emitted, causing ocean acidification. The total radiative forcing is positive and has led to an absorption of energy by the climate system. The largest contribution to the total radiative forcing is due to an increase in the atmospheric concentration of CO_2 since 1750.

Human influence on the climate system is clear. This is evident from the increasing concentrations of greenhouse gases in the atmosphere, from the positive radiative forcing, from the observed warming, and from the understanding of the climate system. Continued greenhouse gas emissions will cause further warming and changes in all components of the climate system. Limiting climate change will require a substantial and prolonged reduction of greenhouse gas emissions over time (IPCC, 2014 p. 33).

Interestingly, at this point in the narrative, there are the approaches taken by the government of the UK and the United States towards the middle of 2000 regarding peak oil and climate change. Hirsch et al. (2005) commissioned by the US Department of Energy suggested to restrain the effects of peak oil by deregulating drilling rules and replacing fossil fuels with other types of fuels, such as biofuels, liquid fuels, bitumen, coal, natural gas, and even nuclear energy (although in terms of environmental damage they are more harmful than oil). On the other hand, The Stern Review (2006) released by the Government of the United Kingdom, in order to mitigate the effects of climate change, proposed the development and use of new technologies, and emphasized the central role of the private sector in creating an international market emission.

The two reports showcase two unidirectional approaches to combat the consequences of climate change and peak oil: both Hirsch and Stern identify solutions to a problem, without considering the other as influential and binding. The reports are antithetical: the concretization of one implies the negation of the other. Both are aimed at finding solutions compatible with today's lifestyle based on a large consumption of energy. Although the premise of both reports may seem noble, in both cases the solutions are compatible and complementary with economic growth, i.e. with the safeguarding of the current global economic system.

The limit was that both proposed a solution-replacement, not a solution-change, so if oil is lacking, we try to replace it with other fuels, if emissions are harmful, we try to replace them with "sustainable" technology. The fossil fuels are both polluting and responsible for the greenhouse effect and limited and according to Chapman (2014), destined to exhaustion. Hopkins (2009) argued that the two problems are not only connected but must be addressed as a single problem. As Latouche and other scholars have argued, the capitalist system is based on the dislocation of resources, production, consumption, and discharge. One civil society instrument, reduced to a mere palliative, is the famous slogan "think global and act local", bait and switch used by governments and multinationals, as Bertorello and Corradi suggested (2011).

This paradox is explained by the fact that the so-called "developed" society is based on the massive production of "decadence", that is, according to Latouche (2007) on a loss of value and a generalized degradation of goods, that the acceleration of "disposable" transforms into waste. This results in men being excluded and fired after use, from presidents and managers to the unemployed, homeless, and other human waste. However, this is the strength of the capitalist system, consciences are lulled into somnambulism and the seas of Africa are inundated with our waste. Bertorello and Corradi (2011) argue that to eliminate nuclear waste we must end the use of atomic energy, and in the same way it is necessary to dispose of the mechanisms of a toxic capitalism, with its ideology of profit and competitiveness, to really eliminate its increasingly poisoned fruit.

The peak of oil instead is simply a hidden problem and silenced by governments and multinational companies. Technology has come to the rescue, in the form of fracking and three-dimensional directional drilling. According to Bardi (2016), US oil production has increased by 54% in just five years, from 3.3 billion barrels in 2009 to 5.1 billion barrels in 2014. Although global oil production rose by only 8.5% at the time, it was enough to prevent it from reaching its peak. Hopkins (2009) argues that this is an illusion, a situation such as the oil crisis of 1973 could be repeated, if the Arab countries drastically increased the price of oil per barrel, causing a strong economic recession.

According to Bardi (2016), changes in temperature and emissions of fossil fuels are unequivocally linked. Moment-by-moment, the change in global average surface temperature of the Earth is proportional to the sum of all carbon emissions at that time. Such carbon-climate response is 1.5° per trillion tons of carbon emitted—equivalent to 0.4 °C per trillion tons of carbon dioxide emitted. Such relationship enabled the negotiators at the recent COP21 meeting in Paris to draft a carbon budget: the world as a whole can emit another 1,000 billion tons of carbon dioxide if the temperature change has to peak at 2 °C, but only another 600 billion tons of carbon dioxide for a roof of 1.5 °C. With current global emissions of around 36 billion tonnes a year, the carbon budget gives the world less than 30 years of business-as-usual emissions if the heating has to reach its maximum at a risky + 2 °C.

The Bet on the Degrowth

We can paraphrase Fotopoulos (2002) in order to synthesize the starting point of Latouche's idea: our economic system is oriented, both objectively and deliberately, towards maximizing economic growth. According to Schumacher, when a society lets itself be dominated by and absorbs such logic, it can only be framed as a growth society, in which growth means producing more, without taking into consequences of this production.

Social security systems, employment, public spending including education, security, justice, transport and health, to mention but a few areas, presuppose a constant increase in the GDP, this also contributes to the creation of an infernal circle, the so-called diabolical carousel known as limitless accumulation; and our future seems premised on such a system. A source of inspiration for the theorist of Degrowth was without doubt Nicholas Georgescu-Roegen and his Bioeconomics theory. It is important to stress that Latouche "briefly" analysed the concept of entropy and of non-reversibility of the transformations of energy and matter.

With his Degrowth theory the French author did not want to follow the path of Georgescu-Roegen and his entropic conception of the production process. The ambitious programme of the 8Rs, synthesis of Degrowth theory, starts its criticism of traditional economics using two main concepts of Georgescu-Roegen's theoretical approach: the theory of production and entropy degradation. It follows the traditional theory of economic growth based on the production function: $Q = Af(K, L, R)$. Suppose that $Q = Af(K, L, R)$ is the quantity of output Q produced by a firm using K units of capital, L units of labour and A technological progress.

However, this function also contains the assumption, taken from the theory of sustainable development, that it is possible to produce any quantity of product, reducing at our will natural resources (R) if the capital stock is sufficiently increased. This is the reason why the classical economicsapproach of substitutability of production factors contradicts the principle of conservation of matter-energy (the first law of thermodynamics), which states that the only difference that can exist between the flow of matter that enters into the economic process and the flow of output waste (produced goods + waste) is qualitative.

It seems clear that is not possible to replace nature in an artificial or infinite way. It is not possible to fill the gap of the inadequacy of natural resources with an increase of capital (R&D, skills, etc.) in order to guarantee the unlimited maintenance of production capacities for the well-being of the mankind. De Benoist (2006) argued that considering natural heritage as capital is already an oxymoron. The value of natural resources is invaluable, and their distribution is priceless, even for the neoclassical approach. There is no substitutable capital for non-renewable resources. At this point, it seems easy to understand how the production of increasing quantities of goods and services implies the use of increasing quantities of raw materials and energy, which consequently implies a massive impact on ecosystems.

The economy thus, ignores the law of entropy, or the irreversibility of the transformations of energy and matter, or in other words, the fact that waste and pollution, even if they are products of economic activity, are not considered to be part of the production process, as this has been determined. Since we eliminated our planet from the production process in the nineteenth century, the last link with nature has been broken. Man transforms resources into waste more quickly than nature is able to turn this waste into new resources. The affirmation of the impossibility of infinite growth in a finite world becomes central for the first time in the economic analysis of Nicholas Georgescu—Roegen who argues the need to replace traditional economic theory with bioeconomy because of the reciprocity of the economic process and the natural environment, the mutual, uninterrupted influence between the two spheres.

To reduce, as Cacciari (2006) posits, the life of a worker to that one of an organism that metabolizes the salary with goods and goods with a salary, from the factory to the hypermarket and from the hypermarket to the factory, three ingredients are indispensable: advertising, credit and accelerated and planned obsolescence of products. Advertising, writes Jean Paul Besset (2006) takes possession of the street, invades the collective space and disfiguring it, takes possession of everything that is public, communication and media, cities, public transport, stations, beaches, and even parties. It floods the night and takes over the day, cannibalizes the Internet, colonizes the newspapers, determines their financial dependence, and makes some of them miserable lackeys. With television, it possesses its weapon of mass destruction, establishing the dictatorship of AI on the main cultural vector of our time. The aggression covers 360 degrees, the hunt is without respite: Mental pollution, visual pollution, sound pollution.

According to Gorz (1992), a survey was conducted among the presidents of the largest American companies. 90% of respondents recognized that it would be impossible to sell a new product without an advertising campaign, 85% claimed that advertising "often" convinces people to buy things that do not they need, and 51% said that advertising convinces people to buy things they do not really want. In order to enable consumption by those who do not have an adequate income and allow the entrepreneurs to invest without having the necessary capital, there is recourse to credit, the increasing indebtedness that provides the means to buy mostly goods that one does not need even when one does not really want to.

It is the shift from "very useful goods" to "goods of great futility" to resume the happy definition of Cacciari. This is, according to Comeliau (2003), the diabolic carousel: the obligation to repay debt with interest and imposing to produce more than we have received. The need to return with interest introduces the need for growth and with it a series of obligations both in the North and, in an even more pressing form, in the South of the world. According to Meadows et al. (2006) all the materials and the energy used by the population and fixed assets do not emerge out of nowhere. They are extracted from our planet, and they do not disappear. When their economic usefulness fails, the scholars argue, the materials are recycled, or they become waste and pollution; energy is dissipated in the form of non-exploitable heat. Flows of materials and energy pass from the planetary sources, through the mediation of the economic subsystem, to the planetary wells that collect waste and pollutants. Finally, through planned obsolescence, the growth society ensures its own survival while the "virtuous" advertising-credit-consumption circle turns into the infernal and suicidal carousel of unlimited growth. In this regard, Latouche (2007) argues that the short timeframes of the life of devices and objects, from electric lamps to glasses, are the desired failure of an element. It is almost impossible to find a spare part or a repairer. However, the repair would cost more than buying the new product (meanwhile manufactured at bargain prices in the galleys of Southeast Asia).

Mountains of computers find themselves in the company of televisions, refrigerators, dishwashers, DVD players, and mobile phones to clog up landfills with various dangers of pollution: more than 150 million computers are transported to the landfills of the Third World with their heavy and toxic metal content (mercury, nickel, cadmium, arsenic, lead). In this type of scenario, the role of the technological innovation becomes central. Herman Daly (1981) defined Faustian as the pact with Great Science and High Technology. He also recognizes the fact that periodic innovation feeds the belief that generated it, the illusion that the new corresponds to the best.

According to Ivan Illich (1992), the aforementioned belief has become an integral part of the modern mentality: every new unit launched on the market creates more needs than meets. It is no wonder that different authors of Degrowth thought to use words like eco-fascism, dictatorship of growth, toxic capitalism, and so on. It is true, such words have a strong impact on us, however, are necessary in order to understand how our current system imposes on developed societies the need to live in a regime of overgrowth.

Capitalism, as Gare (2000) sets out, has reduced humans and natural resources to commodities of exchange, quantifiable values, extending the economic dimension to every sphere of human existence, encouraged by the constant use of increasingly sophisticated technologies and globalization. However, capitalism has united the entire world, shaping it into one large, global market. It has become an "infernal pyramid" which sees at its summit the capitalist economy dictating the rules, with the phenomenon of globalization through technological advances making their implementation possible. Every nation blindly obeys the dictates imposed by global economic policy: increasing production, generating new needs, and stimulating overconsumption.

In his work, Latouche (2007) concludes that, since Technology has become the natural habitat of modern man, it is the latter that must adapt and not vice versa. Technology insinuates itself into and works in our imagination, generating needs and desires to be satisfied and thus influencing the production cycle. Through increasingly efficient means and methods of production, it determines the supply and increases consumption, acting directly on demand. The market is unique and technical innovations across the globe help standardize it, homogenize demand and the desires of the masses. The resources of the planet are exploited by the free market to satisfy almost every country; the technological, industrial and information revolutions have dissolved all geographical borders and linked the whole planet.

Since Les Trente Glorieuses, Piketty (2013) reminds us that capitalism has been fully realized in our society. This is demonstrated by the fact that every nation measures its own well-being through indicators that take only in consideration economic wealth, business relationships, and budget plans in order to pursue infinite growth. What happens when the whole system stops growing? What happens if capitalism can no longer satisfy its ultimate goal?

Over the last ten years, the world economic system has been impacted by a severe crisis brought about by the effects of the Global Financial Crisis (GFC) and the ineffectual regulation of the world's financial systems. Although capitalist economies are no longer able to grow—without irremediably compromising our ecosystem—our society, according to Koch (2013), Navarro and Torres Lo'pez (2014), and Muraca and Schmelzer (2017), influenced by the neoliberal ideology that dominates the economic and political agenda, continues its unbridled climb by mortgaging the future of future generations.

The enslavement of multinational marketing by lobby groups and the international finance system is emptying every single democratic meaning from our society from. The inevitable result is an economic, political, cultural, and environmental crisis. Globalization and its attendant canon of consumerism is homogenizing countries worldwide. Obliterating every shred of heterogeneity, growth policies trace a single line of conduct, destroying any glimmer of diversity that could obstruct their path. What is more, as Castoriadis (2010) argues, the crisis that Western economies are facing, can also be defined as cultural. The unbridled race towards the Western model has destroyed our cultural diversity and our heritage, making us an outcast in our own society.

To elaborate further on this paradoxical situation, according to Castoriadis (2010), man is merely considered a product of this destructive mechanism on which our society is founded. This approach also posits that mankind is destined for programmed obsolescence since people are disposable. Lastly, Castoriadis (2010) suggests the crisis is ecological. The intensive exploitation of the planet's resources, the assiduous annexation of uncontaminated areas, the uncontrolled use of chemical solvents is extremely harmful to soil and atmosphere. The regenerative capacity of the Earth is no longer able to satisfy the demand since man transforms resources into waste more quickly than nature is able to transform this waste into new resources. Man's culture, values and close bond that united them with the earth have been sacrificed on the altar of economic growth.

Capitalism and globalization have bent human heritage and values, not considering the effects of their actions within the cost–benefit calculation. According to Palahniuk (1996), we are living in an historical moment in which an entire generation pumps petrol, serves tables, or is slave within the white-collar system. Advertising makes us pursue the acquisition of machines and clothes, doing jobs we hate, to buy things we do not need. Following Palahniuk's ideas, we are the middle children of history, we have neither a purpose nor a place, we do not have the Great War or the Great Depression. Our Great War is the spiritual one; our Great Depression is our life. We grew up with television that convinced us that one day we would become billionaires, screen legends, or rock stars. However, these dreams have remained unfulfilled, and mankind is slowly realizing it.

Abstractness and conformism have replaced individual identities and their autonomy, the categories of the identical and the different are no longer discernible in our globalized society, which has imposed a single model to be followed. The contemporary man is a different man (some might call him "new") who has lost his spiritual, aesthetic, creative, and contemplative dimension. The fast rhythms of contemporary society impose on him, a frenetic life, no longer suited to the enjoyment of slowness, but dedicated to maximizing production and increasing sales of goods ready to use. The time of knowing has been supplanted by the time of production, even the values of solidarity, trust, and conviviality have been annihilated and replaced by new values based on competition and individualism; the bonds between individuals have been destroyed, thus erasing the exchange of cultural heritage.

The Diabolic Carousel: The Vicious Circle of Growth

Individuality has become a lost mass. It is probably no coincidence that the next step in this infernal mechanism is the realization of Smart Cities, an obsessive dream of academia. Smart City, as Marra (2015) has argued, is the concretization of a communicative interface, a place of interconnection, of Augmented Reality (AR) and real time, of the virtual where the instantaneous transcendence of time and space makes bodies eternal. There, continues Marra, "bodies will fully enjoy the simulacrum of an overflowing and perfect reality. The city would thus turn into the full achievement of an artificial environment, laboratory of anthropological mutation. It is the realm of *smombies*, the neologism coined connecting together the words smartphone and zombie, referring to the wandering metropolitan crowds compulsively isolated and immersed in their phones' activity. That is the new human paradigm: a hyperreal, atomized, unrooted individual endlessly escaping the edge of dematerialization" (p. 89).

The continuous escape forces man to erase the past in favour of an insatiable search for the new. The present is emptied of the gift of meaning and is reduced to an abstract category to plan a future which at the same time is denied to future generations. Man is no longer the guardian of nature; he rather considers himself its master. Subjecting the Earth to his will, man has lost his naturalness; the conquest of territories, even those forbidden, the overcoming of geographical borders, and the forced and continuous withdrawal of renewable and non-renewable resources have made the Earth incapable of accommodating us.

A society of serene degrowth is not feasible if we do not recover those dimensions of life that have been removed by capitalist ideology: the time to participate in the city agenda, the pleasure of free production, artistic or craft, a sense of contemplation, meditation or simply the joy of living. These are what Thierry Paquot (2007) calls "liberated time". It is not about free time, immediately captured by the industries of leisure, health and relaxation, but a reconciliation, often difficult and contradictory, of the individual with himself and nature. The consumerist logic has destroyed the temporal dimension with a simple sentence: time is money. The present is considered an abstract, generic time, to be lived projecting onto tomorrow, it is scarified on the altar of the spasmodic search for the new.

Illich (1992) reminds us that in the current system of programmed obsolescence on a large scale some decision centres impose innovation on the whole society and thus deprive communities of the power of choosing their tomorrow. In this way, it is the instrument that imposes the direction and rhythm of innovation. Latouche (2006) argues the crisis our society is going through is the loss of spatiality and temporality. The annulment of human values is the dramatic consequences of the decision to vote for growth as an end in itself. Since the advent of the capitalist era, there have been numerous instances of criticisms of its modus operandi, of human and environmental degradation, and of the growing induced gap within society and between different societies.

The cornerstone of the literature within the Degrowth paradigm is Marxist criticism. The is approach has analysed the contradictions of nascent capitalism, highlighting the social unease and alienation of workers, forced to undertake degrading work to increase the profits of the capitalists. According to Marx (1964), the critique of capitalism is class struggle. It is the revenge of those who are exploited by the new logic of the market, and who are able of revolutionizing modern society, declaring the end of capitalism and the beginning of a new, socialist society.

Capitalism is still the dominant ideology and has found new life through the phenomenon of globalization and the recent information revolution, which have allowed the economy and technology to advance and expand into every corner of the globe, with maximum efficiency, speed and power, persuading each country and each individual to obey its rules.

Although the risks of a politics devoted to productivity and consumerism are evident and well known, decision makers, such as lobbyists and politicians, deny the illness and crises they are experiencing, demanding greater sacrifices from the world population in order to preserve and revive the capitalist model. Crisis is considered by the proponents of the capitalist model to be a temporary instability of a complex and dynamic system. However, for the growth objectors since the days of Illich "left small masonry" (Latouche 2004), crisis is the first sign of a point of no return: unless there is a change in direction, human, spatial, and temporal conditions will worsen and lead world society to oblivion.

Modern Heroes

The crisis is therefore framed by Degrowth scholars as an opportunity for a radical change. It is an opportunity to rebalance the relationship between man and nature, re-establishing the link with Earth and rediscovering dialogue with others, breaking down the barriers created by economics, the technological age, and globalization. Degrowth is a valuable example of voluntary simplicity. It is not a unique model of life unconditionally applicable worldwide in every country in the same way, but rather an economic, philosophical, and political alternative to be realized in a concrete, autonomous and diversified way in the various territorial realities. The ways to achieve it are unique and unique to each nation.

Let us remember that Latouche, pioneer of the Degrowth paradigm is, above all, an economist. That means that "his" Degrowth is substantially an alternative economic approach. Degrowth wants to break with the liberal paradigm and to overthrow the logic of growth itself. Degrowth wants to consider Earth as a fundamental subject of the equation within the production processes, respecting its limits. Moreover, Degrowth is a new political vision, a laboratory of inclusive democracy, a new, locally-based way of government that allows e direct participation of citizens in the decisions of governments, an awakening the social conscience and a rehabilitation of the duty of citizens to the care of their own community within their respective country. In addition, Degrowth is a *weltanschauung*. It is a real mental shift in the collective imagination of contemporary man, an awakening from the torpor induced by the growth society. Man becomes a new man, *homo resaliens*, who has the artist's creative soul, the reflective spirit of the philosopher, and the revolutionary nature of the growth objector. Embracing Degrowth is an opportunity for mankind to rediscover a renewed aesthetic sense, enjoy the slowness and frugality of life and regain possession of their heritage, of their traditions and know-how. The rediscovery of genuine values, such as solidarity, conviviality, and participation, will allow homo resaliens, inhabitant of a Degrowth society based, to live in harmony with the *alter*, building relationships through a renewed language that respects the singularities, the culture of each one.

Degrowth therefore could be framed as a form of positive globalization in which the rediscovered values will help to re-establish the identities of each territorial reality. Frugality proposed by Degrowth will be the weapon used in fighting planned obsolescence of goods and the diabolical carousel of the overabundance. The sense of the finite is the perfect tool to use against the decision makers, who for too many years have been demanding austerity, sacrifices in order to reach the unsustainable well-having, deliberately confused with the well-being, a fundamental requirement to help a reconstruction of a long-forgotten relationship with nature.

According to Gare (2000, p. 40) in order to "create and develop this new culture a far greater effort will be required to develop the potentialities of people than currently occurs within a capitalist society". This will include, Gare argues, the creation of educational institutions, from kindergartens to universities, "which socialize people into a culture of creativity and sensitivity in which all people will become simultaneously workers, horticulturists, engineers, poets [...], and will take the development of people's potentialities to participate in this culture of creativity as the ultimate end of society".

Only by following this path, the philosopher continues, "will such eco-socialist, socio-economic forms be able to survive, challenge, prevail over and then subordinate the social mechanisms of the capitalist mode of production". The past will be rehabilitated, memory, and history will be fundamental to reconstructing the present, according to the wisdom of the elders, looking to a future that allows to increase the potential of man, respecting the planet. Time will follow circular dynamics, abandoning the linearity marked by capitalism.

Rewriting the literature twice is not the first aim of this manuscript, and, in addition, this research path does not want to replicate Latouche's work.

This research aims to address the criticism which emerged ten years after the publication of the *Petit traité de la décroissance sereine* (2004) i.e. that lack of concrete examples, that the French author replaces with digressions on austerity, downshifting, and voluntary simplicity, all surrounded by a moralistic attitude regarding the principles of responsibility and sense of limitation. In the thinking of Latouche everything is simplified, focused on the change of our collective imagination, (perhaps due to the holistic formation of the French heritage) that fails to break down the individual issues and apply his Degrowth on a real field of analysis.

Although Latouche recognizes the interdependence of the elements of a complex economy, such as the current one, in which producers, money, consumers, goods, and the environment interact with each other; the allocation of resources is barely touched on, which, spared in spending that degrades the environment, can stimulate the production of "relational goods" if possible, out of the market. Consumption and production are in fact closely related to each other but focussing only on the change of the imaginary without analysing the economic impact of concepts such as self-production, non-mercantile goods, off-market transactions is limiting for the purposes of investigation and resolution.

The Re-establishment of Politics

Latouche lingers on the self-production of yogurt; however, such examples do not contribute to answer the dilemma of Georgescu-Roegen's *homo exosomatic*: is mankind disposed to consider a programme that implies a limitation of its addiction to exosomatic comfort? After more than ten years of data analysed by the CSIRO, WWF, and so on, objective growth without focusing attention on the economic and material conditions that cause unsustainable pressures on the environment, society and ecosystem response (resilience or impact resistance) would be an end in itself. Today's humanity already consumes 30% more of the biosphere's regeneration capacity. Clearly, this average "ecological footprint" hides huge disparities. From the data provided by Global Footprint Network, each Australian consumes 6.9 hectares per person (the country with the 6th highest Ecological Footprint in the world).

Most African countries consume less than 0.2 hectares per capita of bio-productive space, which corresponds to one-tenth of the planet and at the same time provide us with food for our livestock. In addition, in order to conserve biodiversity and ensure the survival of species, especially wild species, it is necessary, according to Jean Paul Besset (2006) to share space with other species, leaving them with the last 20% of the land space. However, for humanity this is still not appropriate. This is due to the interruption of the systematic character of the process of territorial transformation, of the creation of infrastructures, and of urbanization.

For many geopolitics, however, the solution of the ecological footprint has seemed, so to speak, almost trivial in its ease. In fact, it is sufficient to shift the perspective from the limited availability of natural resources to the reduction of the world population. It is a simple matter of common sense: a finite planet cannot sustain an infinite population. Analysts and the industrial world persist in trying to reconcile two contradictory principles such as growth and respect for the environment, proposing a scenario based on three aspects: the substitutability of factors, the knowledge economy, and eco-efficiency. The first law of thermodynamics denies the feasibility of the first aspect, despite this replacement being the key factor supporting technological progress even if resources become increasingly scarce, as well as the basis of development theory sustainable. We do remember the rejection of Robert Solow against *Limits to Growth*. The Nobel Prize winner in 1974 argued that the world can in fact, go on without natural resources.

The most recent *transformism*, the most captivating of our system is the one called capitalism of knowledge (or knowledge economy), founded on immaterial production. However, the intangible goods are produced by the harsh materiality of the work. The "knowledge economy" is drastically demystified by Paul Krugman (2011), who states that computers excel in undertaking routine tasks. It follows that any routine task, a category in which many non-manual white-collar jobs are included, is exposed and vulnerable. On the other hand, jobs that cannot be completed by following explicit rules, a category in which many types of manual labour fall, from truck driving to caretaking, will tend to increase, even in full technological progress. The economist continues that most of the manual work that has yet to be carried out in our economies seems to be of a kind that is difficult to automate.

Meanwhile, much of the white-collar work done by well-educated and relatively well-paid workers could soon be computerized, according to Krugman (2011). The Nobel Prize winner shares Latouche's vision, i.e. what we need is a shared welfare society, education is not the answer: what we must do is to create that new society. We must be able to recover the bargaining power that the workforce has lost over the last thirty years, so that both ordinary workers and the super-capable have the power to negotiate a good salary. At this point, we are facing a bottleneck: if industrial productivity is in deep crisis and the knowledge economy already sees threatening clouds on its horizon, whose is the future?

The third point, ecological efficiency, based its belief in that technological progress will allow to produce increasing quantities of goods with an even smaller use of matter and energy. The combination of technological progress and growth make it, at the same time, the only solution to the ecological crisis by reducing the impact on ecosystems through greater efficiency in the use of resources (eco-efficiency), naturally after the economies have achieved a certain level of development. Here we collide with the harsh reality, the reality preconized by William Stanley Jevons. The neoclassical economists believed that technological progress developing efficient power sources would decrease energy consumption.

He observed how a reduction in the limits of use of a technology is associated with an increase in consumption linked to it, making it possible to understand how each time we manage to save energy or raw materials to manufacture a product through more efficient technologies, the positive effect of this gain (saving) is over-compensated by the incitement to consume resulting from it and by the increase in the quantities produced.

Subsequently, the consumer society, framed as a representation of technical progress, stimulates new needs and, with the combined effect of marketing and advertising, induces a profound transformation in the habits of life and consumption. It is this transformation that fuels the increase in consumption, which more than compensates for the reduction in the use of resources linked to greater efficiency, leading to an increase in the use of raw materials and a greater impact on ecosystems.

Overpopulation, overabundance, and the technological paradox can be framed as direct causes of ecological risk. In particular, according to Commoner (1986), technology is the critical link between profits and pollution, through increases in productivity, but at the same time damage to the environment. The Rebound Effect (a way to understand the Jevons paradox) and the ecological footprint are two notions extremely important for understanding the paradigm of Degrowth promoted by Latouche. In addition, what has been said about environmental degradation brings out one of the distinctive features of the paradigm of the French author: not merely denouncing the logic of the dominant capitalist system but extending its criticism to every doctrine that finds its foundation in the development of productive forces.

The alternative political project of post development is largely identical to that which in the eighties was defined as "ecological rationalization" by Andrè Gorz, who coined the motto "less is better". Its purpose is a society in which one will live better by working and consuming less. The ecological modernization imposes that the investment no longer serves growth but Degrowth of the economy, that is to a narrowing of the sphere governed by economic rationality in the modern sense. According to Gorz (1992) there can be no ecological modernization without restriction of the dynamics of capitalist accumulation and without reduction of consumption for self-limitation. The essential condition for making the reduction in consumption acceptable and understandable to all is that collective wealth is directed towards socially useful activities with a low ecological impact.

The GDP of Happiness

Reducing consumption means saving energy through innovative solutions and a different use of available resources. We must not produce more; on the contrary, it is better to divide the work among young people, women, the elderly, and migrants and ensure that every life regains the time "freed" from work in which to savour the slowness and have the opportunity to find themselves and others. This defines the largest ecological paradox of growth, a direct consequence of the erroneous equation according to which individual well-being ("happiness") is measurable through consumption and the quantity of goods that can be purchased.

GDP is an imperfect gauge, GDP is, according to Latouche (2005), nothing but a flow of purely mercantile and monetary wealth; growth is its progression and therefore everything that can be sold and that has a monetary added value, contributes to inflating GDP and growth, regardless of whether this contributes to collective individual well-being or not. This indicator does not include many activities and resources that contribute to well-being, simply because they are not of a mercantile nature or because they do not have a direct monetary production cost such as the impoverishment resulting from the exhaustion of natural resources, or according to Gadrey and Jany-Catrice (2005), the freed time and relational goods.

As the philosopher Alain Badiou (2006) observes, much of contemporary oppression is about time. We are forced to fragmented, discontinuous, dispersive times, in which speed is the main factor. It is not the time of the future project, but the one of the consumptions, of wage earners. Courage could consist in trying to impose another timing, another rhythm. The future is not in industry and technology, even though they will continue to be important, but, according to Krugman (2011), the future lies in agriculture. Bevilacqua (2011) reminds us of the social, economic, and environmental functions carried out for centuries by agriculture. According to the author, the peasants produced, not only agricultural goods, but sheltered walls, reactivated paths, channelled rainwater, cleaned up stains and woods, planted trees, treated soil health, etc.

Nowadays, this immense and age-old work of care and control is no longer carried out by anyone, except for the sparse interventions of some mountain communities. The most evident and dramatic sign of this abandonment is the destructive frequency of the fires. These events, in addition to destroying huge amounts of biomass, killing people and animals, striking biodiversity oases, even destroying trees and valuable vegetation, have had pernicious effects on the ground. The chemical alteration produced by the fire, in fact, "cooks" the earth, makes it bare and fragile, exposed to atmospheric phenomena, so that the violent autumn rains often drag it ruinously downstream, giving rise to landslides.

It is no coincidence, according to Kempf (2010) that the labour movement has collapsed, on the other hand, most of the emblematic struggles of our time concerns the problem of genetically modified organisms and is largely supported by the peasant movements. It is a matter of preventing an industrial model based on environmental disregard, on the scarce use and on the appropriation of living resources with patents to be imposed on peasant agriculture. In recent times, agriculture has begun to turn into a project full of ecological and social values and meanings, applying itself as a lever for the rebirth of local economies, with the offer of good opportunities for stable employment, especially if the share of industrial automation of many operations is reduced.

Providing a Concrete Alternative: The Transition Towards the Post-development

Starting from the criticism as voiced by Jeroen van den Bergh (2011) and analysed in the previous chapter related to the literature and taking inspiration by the paper "*In defence of degrowth*" (Kallis, 2011), the aim of this paragraph is the concretization of a new narrative. Mankind, one of the many species until last century, has become a geological agent capable of transforming the structure of the biosphere: the concentration of carbon dioxide in the atmosphere is the highest it has ever been there for about a million years, more than 40,000 artificial dams heavily modify natural sedimentation, the acidity of the oceans increases rapidly while the disappearance of living species is comparable to that which destroyed the dinosaurs at the end of the Cretaceous period.

Hervé Kempf (2010) argued that we need to start a new era, a new society, unless we prefer the fate of the dinosaurs. Degrowth theory has its grassroots in the academic field and develops as an ideological challenge to the dominant economy, i.e. ultraliberalism. The slogan of "degrowth" appeared as a "semantic bomb" or an "explosive term" quoting Paul Aries, capable of breaking the weak consensus of submission to the dominant productivist order, or in other words of initiating a decolonization of the imaginary. Castoriadis maintains we need: "a new creation of imaginary which has an unparalleled importance in the past, a creation that would put at the centre of human life a more meaningful rather than the expansion of production and consumption, which would indicate the different life goals, in which it would be that would be recognized by human beings as goals worth living for. This is the immense difficulty we are called to face. We should want a society in which economic values have ceased to be central (or unique), where the economy is put back into its place as a mere means of human life and not as its ultimate end, in which it waives this mad rush towards the indefinite growth of consumption. All this is not only necessary to avoid the definitive destruction of the terrestrial environment, but also and above all to free ourselves from the psychic and moral misery of our contemporaries" (Castoriadis, 1996, p. 96). However, the Greek-French philosopher continues:

> in order for such a revolution to take place, it is necessary that profound changes take place in the psychosocial organization of Western man, with respect to his attitude towards life, in his imagination. It is necessary that the idea that the only purpose of life is to produce and consume—an idea both absurd and degrading—is abandoned; it is necessary that the capitalist imagination of a pseudo-rational pseudo-control, of an unlimited expansion, be abandoned. Only men and women can achieve all this. A single man or a single organization can at most prepare, criticize, incite, sketch out possible orientations. (Castoriadis, 2005, p. 244)

However, to arrive at a solution to get out of the dominant imaginary, we need to understand how we have entered this process of economization of the spirits and commodification of the world. According to Castoriadis (1975), economics is a mere invention (*The Imaginary Institution of Society*). This thought is strengthened in his contribution to the book of Mendes *Le mythe du développement*, 1977 in which the two authors deal with the crisis of development framed as a crisis of the corresponding imaginary meanings and in particular of progress. The incredible ideological resilience of development is based on the no less surprising resilience of progress: "an increasing number of people do not believe in progress. Everyone wants to acquire something more for next year, but no one believes that the well-being of humanity resides in the growth of 3% per year of the level of consumption. The imaginary of growth is certainly always present and is also the only one to survive in the Western world. Western man no longer believes in anything except the fact that he will soon be able to have a high-definition TV" (Castoriadis 2005, p. 220).

That is exactly what is still preventing a large number of people from joining Degrowth philosophy. How can we escape from the state of drug addiction of the consumer society? Certainly, this is the junction of the problem. It is not possible according to Latouche (2013), because it is not possible to decide to change one's own imagination and even less that of others, especially if these are dependent on the growth drug. However, let us think, first of all, about the *paideïa,* which plays an essential role for Castoriadis.

A New Humanism

The *paideïa*, the education of citizen, according to Castoriadis (2012), means the opportunity for citizens to participate, the fact of standing against the anonymity of a mass democracy. Nobody is born a citizen, continues Castoriadis, you learn it, by observing the city you are in, and not watching television as you do nowadays. Unfortunately, this remedy is not fully possible except on condition that the degrowth society is already realized. According to Latouche (2013) to achieve such, it should be out of the consumer society and its "civic stupidity" regime. The denunciation of advertising aggression, a vehicle of today's ideology, is certainly the starting point of the counteroffensive to emerge from what Castoriadis calls "consumerist and television onanism".

If the problems of development/underdevelopment and of North–South relations are not at the cornerstones of Castoriadis's reflections, however, it is not so for postcolonial anthropology, which is probably the second most or less conscious source of the invention of the expression of decolonization of the imaginary. In 1969, Gérard Althabe titled his studies on Madagascar *Oppression et libération dans l'imaginaire* (*Oppression and liberation in the imaginary*), however, the turning point was in 1988, the year in which Serge Gruzinski titled his book *La colonisation de l'imaginaire* (*The colonization of the imaginary*). However, as Latouche

(2013) argued, when Gruzinski speaks of the colonization of the imaginary, it is still a continuation of the colonial process in the strict sense and, eventually, of the conversion of the natives by the missionaries.

The change of religion constitutes, according to Latouche, at the same time a deculturation of the spirits and an acculturation to Christianity and Western civilization in the framework of the imperialist project. Growth and development are inextricably linked to the process of conversion of mentalities, therefore with a process of an ideological (almost religious) nature, tending to establish the imaginary of progress and economy, however, in accordance with the definition of Aminata Traoré (2002) the *violation of the imaginary* it remains symbolic. With the colonization of the imaginary in the West, we find ourselves struggling with a mental invasion of which we are the victims but also those responsible. It is no coincidence that Serge Latouche (2013) has always talked about self-colonization or voluntary servitude.

At this point, it is a matter of operationalizing the inverse procedure, that is, decolonization of the imaginary, the heart of the French professor's theory (largely the object of study of this research). According to Latouche (1989), the West was effectively assimilated into a "deterritorialized" paradigm characterized, on the one hand, by a belief, unheard of if carried on the scale of the cosmos or of cultures, in a cumulative and linear time and by the attribution to man of the mission of absolute dominator of nature and, on the other hand, in the cult of calculating reason considered capable of organizing its action. In order to understand this paradigm, Latouche used (1989) the ten-element codification proposed by Johan Galtung:

Characteristic features of the western social cosmology:

- linear time conception, centred on the present.
- An eminently analytic, and not holistic, conception of epistemology.
- Conception of human relations in terms of domination.

Characteristic features of the western social structures:

- division of vertical and centralized work.
- Conditioning of the periphery by the centre.
- Marginalization: social division between inside and outside.
- Fragmentation: atomization of individuals within groups.
- Segmentation: division between individuals.

If these are the premises, it is easy to understand that a decolonization of the imaginary constitutes a real revolution. It is not only a cultural revolution, it is about escaping from the economy, changing values and then de-westernizing ourselves. Degrowth theory finds its expression in the aftermath of 2007 with the explosion of the financial bubble, the so-called subprime crisis and with the subsequent collapse of the world's major banks, the Lehman Brothers financial organization. However, the theoretical reflection on post-development, which made its way to France and almost made its way in the underground between 1972 (the era of Nicholas Georgescu-Roegen, Ivan Illich and André Gorz) and 2002 with the UNESCO convention *Défaire le développement, refaire le monde*, advocates the crisis of society in the globalized market.

It also proposes a (positive) way out of well known and autonomous democratic and ecological societies, societies based on the degrowth. The financial crisis of 2007/2008 predicted in 2005 by Rajan (2005), is a crisis of civility, a cultural and social crisis; it is not just financial/economic crisis. The substantial difference between the growth objectors (or Illich's left small freemasonry) and alter-globalization scholars, for instance the solidarity economy, is that the scholars of post-development do not frame the core of the problem in the terms of neoliberalism, what Karl Polanyi (2001) calls the formal economy, but in the logic of growth perceived as the essence of economics. The difference is radical. This is not a matter of substitution, for example replacing a "bad" economy with a "good" one. There is no good development, it is not about building a model of positive growth. Nor is it a question of creating a hybrid system that combines liberalism with intervention of the state and of the gift paradigm.

Beyond the Currency

What the growth objectors proposed was to get out of the current economy. According to Kallis (2011), we should consider Degrowth as "a growth" in the same way we refer to atheism, and that is exactly what it is: becoming atheists of growth and economy. Every human society has to organize the production of its life, they have to reasonably use the resources of its environment and consume them in the form of material goods and services, that is the path of any Degrowth society as well, rather like the societies of abundance of the stone age, described by Marshall Sahlins (1972), which have never entered the economy. This society does not take its first steps in the iron grip of scarcity, of needs, of the cold economic calculation made by the mind of *homo oeconomicus*.

The same imaginary foundations of the institution of the economy must be called into question. The rediscovered frugality makes it possible to rebuild a society of abundance based on what Ivan Illich called "modern subsistence" i.e. the lifestyle within a post-industrial economy in which people have managed to reduce their dependence on the market, and have come to protect—by political means—an infrastructure in which techniques and tools serve as a priority to create non-quantified and non-quantifiable use values of the owners of goods production (Illich, 2005).

Latouche (2006, 2007, 2013) asserts that a degrowth society can invent itself from the current situation, but also, at worst, from the ashes of the consumer society. It will not necessarily eliminate money, markets, or wage labour. This does not mean that it will be a society dominated by money, a market society, and wage labour. Any degrowth society will fail (and it is not even its primary objective) to abolish private ownership of the means of production and even capitalism, (at least in the beginning); however, it will succeed in stemming capitalism and its thirst for growth. The transition towards *DemocraCity* implies forms of regulation and hybridization. If moral conviction excludes the compromise of thought, the ethics of responsibility presupposes compromise. Thus, according to Latouche (2013), concrete proposals by the alter-globalists and supporters of the solidarity economy are welcome. This is the reason Latouche (2013), supports the 8Rs of the concrete utopia, and the 10 Rs of a political nature in which alternative movements can find space and a common territory:

1. Rediscover a sustainable ecological footprint.
2. Reduce transport by internalizing costs through appropriate eco-taxes.
3. Relocate activities.
4. Reconstitute peasant agriculture.
5. Reassign productivity bonuses by reducing work time creating new employment.
6. Relaunch the "production" of relational goods.
7. Reduce energy waste by a factor of 4.
8. Reduce the advertising space.
9. Reorient the techno-scientific research.
10. Regain possession of money.

Thanks to the last item of the programme, it is possible to find the link between Degrowth and resilience and to investigate transition cities as a materialization of the Latouchian theory. The concept of resilience is defined by the French author as more explanatory and rigorous than that of self-sustainability. Due to the concept of resilience, it seems possible to understand the last item of the political agenda proposed by Latouche. According to the French economist, we must gradually regain possession of money and no longer leave it exclusively in the hands of the banks and, if possible, even deprive them of all the monetary properties. According to the anthropologist David Graeber (2011), money has to serve, not to enslave. Latouche argues (2013) that it is time to invent a real local monetary policy.

In order to maintain the purchasing power of the citizens, monetary flows should remain as much as possible within the region. Economic decisions should be taken at the same level. Lietaer (2006, p. 76) as one of the inventors of the Euro acknowledged: “Encouraging local or regional development while maintaining the monopoly of the national currency is tantamount to trying to detoxify an alcoholic with gin”. According to Lietaer and Kennedy (2008) to this it is appropriate to define the boundaries of banking and finance, to reconfigure the global financial market and to re-fragment monetary spaces.

The development of alternative, local, bio-regional, complementary currencies (with different formulas to be tested and adapted: rotary mortgage credit, negative interest rate, etc.) participates in this objective, but is also a powerful lever for re-locating, that is, to reappropriate one’s own territory of life and to rehabilitate the world, as opposed to de-territorialization, to non-places and to the de-temporalization of globalized productivism. Taking possession of money also means regaining a certain mastery of time, thus loosening the grip in which the obsession of its price imprisons our lives.

According to Desmonde, as reported by the aforementioned Lietaer and Kennedy (2008), the primitive coin symbolized the reciprocity between people connecting them emotionally with their community. The coin was originally a symbol of their soul. Latouche argues (2007) that a reasonable scale for a regional monetary system undoubtedly in a range between 10,000 and 1 million people. This corresponds to a “bio or eco region” and represents a balance between efficiency and resilience. Efficiency involves centralization in order to benefit from economies of scale (but with the risk of fragility due to mono-functionality and hyper-specialization) where resilience (i.e. the ability to adapt to change) supposes the small scale and the multi-functionality. The diversity necessary for the resilience of ecosystems (natural or human) therefore presupposes a certain fragmentation of spaces.

An Historical Background of the Transition Town Movement

The Transition Town Movement born in Ireland, in Kinsale in 2004, is perhaps the form of construction, from below (bottom-up), of what can most closely approach a society based on Degrowth. Rob Hopkins (2008) proposes to his students, to imagine a scenario of Kinsale facing the peak of oil scenario and to propose solutions based on the principles of permaculture. In response his students produced the Kinsale Energy Descent Action Plan, a document with guidelines on how to reduce energy consumption and optimize resources, for the use of citizens and institutions. This document will become the cornerstone of the Transition movement. Hopkins, in fact, returns to the place of his origin, the city of Totnes, in England, where he starts his path towards transition. These cities, according to the Hopkins (2008), tend to primarily prioritise energy self-sufficiency in anticipation of the end of fossil energies and more generally resilience.

This concept, inspired by scientific ecology, can be defined as the ability of an ecosystem to resist changes in its environment and to regain new life after a shock. How can large urban centres prepare for the end of oil, rising temperatures, and all foreseeable catastrophes? The answer coming from ecological experience is that, if specialization allows to increase the performances in a sector, it weakens the resilience of the whole. Vice versa, diversity strengthens resistance and adaptability. The reintroduction of gardens, proximity agriculture, small handcraft realities, the multiplication of renewable energy resources reinforce resilience.

According to Hopkins (2008), Transition Towns is based on six fundamental principles:

1. **Visioning**: be able to clearly imagine realistic and desirable goals in the long term.
2. **Inclusion**: the ability to remove the barriers between social areas, no more “them” or “other” but just “we” that create links and dialogue between the various social groups.

3. **Awareness-rising**: the need to properly communicate peak oil and climate change information to people, especially given the constant and possibly less and informative media messages. On the one hand there are reports that are consistent with the current situation, such as energy wars and social protests, on the other it is reiterated that encouraging consumption in the global free market is the only engine of development, calculated by applying indicators of GDP. This accords with the position outlined by Latouche (2007) summarized in his "diabolical carousel".
4. **Resilience**: the ability of a substance or object to spring back into shape, in this case, the ability of the community to sustain itself following severe external shocks, such as a crisis situation.
5. **Psychological insight**: supporting, through groups and the community, people on the path of raising awareness so that they do not feel isolated and powerless but an active and interdependent part of a collective intervention.
6. **Credible and appropriate solutions**: problems must be faced by proposing realistic and desirable solutions, outcomes of a positive vision. Although the oil crisis and climate change remain the starting points of the transition, these problems (which can be defined as catastrophes) should not occupy the centre of the discussion; they should simply be the starting point of any discussion to arrive at possible and desirable solutions. It is neither desirable nor possible for the individual to take responsibility for the whole planet. The individual will not make the difference by lowering the thermostat or by taking only two showers a week. Approaches of this type will only generate frustration for the individual and a misplaced trust in government policies. On the other hand, approaches of the individual with expressions that invoke government intervention, make the citizen distant, without counting the disaffection to politics (because it will never be involved in decisions at any level). A sense of belonging, of being part of a community makes it possible for the individual to be motivated, to show active commitment and participation in order to achieve a change set in the collective imaginary (and a concrete change within the society).

Paraphrasing Richard Heinberg (2015), the future will bring a permanent recession and we probably will not be able to grow the economy using less energy. We may have to learn to get by without economic

> growth, something that no politician would like to hear, let alone admit. The problem is that we give priority to economic growth over everything else. It may be time, as Heinberg argues, to measure the progress of civilization in some other way, perhaps using something along the lines of the Genuine Progress Indicator or an indicator such as the often-mocked Bhutan Indicator of Gross National Happiness.

Virtuous examples of Degrowth, valuable example of the new civilization envisaged by Gare (2002), Latoucheand Kallis (2011), Jackson (2013), have arisen all over the world. One example is the reality of Transition Darebin, located in Melbourne. Transition Darebin is used as a case study in this research to demonstrate how communities based on principles of solidarity, mutual aid, virtuous networks, and local resilience initiatives can, starting from the bottom, influence, change and ultimately bring about the end of old-growth systems.

Growth society has shown its contradictions and its inherent danger, embracing the Degrowth alternative could be a desirable answer to end short-sighted anthropocentrism, the plundering of nature, and the slaughter of animal species. A new morality, core for the concretion of *DemocraCity*, (explained in detail in Chapters 5 and 7), requires a redefinition of the relationship between man and nature.

Anthropocentric values are no longer exclusive but are combined with those of the living; and the living is all that is endowed with life including the biosphere, with the vulnerable richness of its species. In conclusion, using the lesson of Pascal contained in his *Pensées*, we can argue that Degrowth is a "bet" in which the choice between the existing system and one based on a different logic is governed by the disproportion of the stakes:

- If the goals of Degrowth theory are achievable, and I believed in the theory and took part in their realization: I won.
- If, however, the goals of Degrowth theory are not achievable and I believed and took part in their realization: I have not lost or won.
- If Degrowth is achievable, and I did not believe in it: I lost.
- If Degrowth is not achievable and I neither believed nor took part in it: I have not lost or won.

The road towards *DemocraCity*, the first step towards a society of frugal abundance, will preserve the environment that is ultimately the basis of all life, will open to everyone a more democratic access to the economy, reduce unemployment, strengthen participation (and therefore integration) and also solidarity, and strengthen the health of citizens due to the growth of sobriety and the reduction of stress.

References

Ali, M. (2013). *Climate change impacts on plant biomass Growth*. Springer.

Badiou, A. (2006). *Il secolo*. Feltrinelli.

Bardi, U. (2007). Energy prices and resource depletion: Lessons from the case of whaling in the nineteenth century. *Energy Sources*, Part B, *2*(3), 297–304.

Bardi, U. (2016). What future for the anthropocene? A biophysical interpretation. *Biophysical Economics and Resource Quality, 1*(1), 1–7.

Bardi, U., & Lavacchi, A. (2009). A simple interpretation of Hubbert's model of resource exploitation. *Energies, 2*(3), 646–661.

Bertorello, M., & Corradi, D. (2011). *Capitalismo tossico. Crisi della competizione e modelli alternativi*. Edizioni Alegre.

Besset, J.-P. (2006). *La scelta difficile. Come salvarsi dal progresso senza essere reazionari*. Dedalo.

Cacciari, P. (2006). *Pensare la decrescita. Sostenibilità ed equità*. Intra Moenia.

Castoriadis, C. (1975). *L'Institution imaginaire de la société*. Seuil.

Castoriadis, C. (1996). *La montée de l'insignifiance* (p. 96). Les carrefours du labyrinthe IV.

Castoriadis, C. (2005). *Une sociétè à la dérive* (p. 244). Seuil.

Castoriadis, C. (2010). *Relativismo e democrazia. Dibattito con il MAUSS*. Elèuthera.

Chapman, I. (2014). The end of peak oil? Why this topic is still relevant despite recent denials. *Energy Policy, 64*, 93–101.

Comeliau, C. (2003). *Brouillons pour l'avenir: contributions au débats sur les alternatives, Iued*. Puf.

Daly H, 1981, *Lo stato stazionario. L'economia dell'equilibrio biofisico e della crescita morale, Sansoni*. Firenze.

De Benoist, A. (2006). *Ultimo anno. Diario di fine secolo*. Edizioni Settecolori.

Gadrey, J., & Jany-Catrice, F. (2005). *Les nouveaux indicateurs de richesse*. La Decouverte, Reperes edition.

Gare, A. (2000). Creating an ecological socialist future. *Capitalism Nature Socialism, 11*(2), 23–40.

Gare, A. (2002). Towards an inclusive democracy: The crisis of the growth economy and the need for a new liberatory project, by Takis Fotopoulos. *Review of Radical Political Economics, 34*(1) (34), 97–99.

Gorz, A. (1992). *Capitalismo, socialismo, ecologia*. Manifestolibri.

Graeber, D. (2011). *Debt: The first 5,000 years*. Melville House.

Hertzberg, M., Siddons, A., & Schreuder, H. (2017). Role of greenhouse gases in climate change. *Energy & Environment, 28*(4), 530–539.

Hirsch, R. L., Bezdek, R., & Wendling, R. (2005, February). *Peaking of world oil production: Impacts, mitigation, & risk management*. Science Applications International Corporation, U.S. Department of Energy, National Energy Technology Laboratory.

Hopkins, R. (2008). *The transition handbook: From oil dependency to local resilience*. Chelsea Green Publishing.

Hopkins, R. (2009). *Manuale pratico della transizione: Dalla dipendenza dal petrolio alla forza delle comunità locali*. Arianna.

Hubbert, M. K. (1949). Energy from Fossil Fuels. *American Association for the Advancement of Science, 109*(2823), 103–109.

Illich, I. (1992). *La convivialità*. Red, Como.

Illich, I. (2005). *La convivialità, Una proposta libertaria per una politica dei limiti allo sviluppo*. Boroli.

IPCC. (2014). Climate change 2014: Synthesis report. Contribution of working groups I, II and III to the fifth assessment report of the intergovernmental panel on climate change [Core writing team, R. K. Pachauri and L. A. Meyer (eds.)]. IPCC, Geneva, Switzerland, 151.

Jackson, T. (2013). New economic model needed not relentless consumer demand. *The Guardian Blog*.

Kallis, G. (2011). In defence of degrowth. *Ecological Economics, 70*, 873–880.

Karl, T. R., & Trenberth, K. E. (2003). Modern global climate change. *Science, 302*, 1719–1723.

Kempf, H. (2010). *Per salvare il pianeta dobbiamo farla finita con il capitalismo*. Garzanti, Varese.

Koch, M. (2013). Welfare after growth: Theoretical discussion and policy implications. *International Journal of Social Quality, 3*(1), 4–20.

Krugman, P. (2011). Degrees and dollars. *New York Times*.

Latouche, S. (1979). *Critique de l'impérialisme. Une analyse marxiste non léniniste de l'Impérialisme*. Anthropos, Paris.

Latouche, S. (2004). *Decolonizzare l'immaginario. Il pensiero creativo contro l'economia dell'assurdo*. EMI.

Latouche, S. (2006). *The globe downshifted*. Le monde diplomatique.

Latouche, S. (2007). *La scommessa della decrescita*. Feltrinelli.

Latouche, S. (2011). *Come si esce dalla società dei consumi. Corsi e percorsi della decrescita*. Bollati Boringhieri.

Latouche, S. (2012). Can the Left Escape Economism? *Capitalism Nature Socialism, 23*(1), 74–78.

Latouche, S. (2013). Prefazione in Bonaiuti, M *La grande transizione: Dal declino alla società della decrescita*. Bollati Boringhieri, Torino.

Latouche, S. (2016). *La decrescita prima della decrescita. Precursori e compagni di strada*. Bollati Boringhieri.
Lietaer, B. (2005/2006, Rapport). *Des monnaies pour les communautés et les régions biogéographiques: un outil déci- sif pour la redynamisation au XXIème siècle*, in Blanc J., *Exclusion et liens Financiers, Monnaies sociales, Economica*, 76.
Lietaer, B., & Kennedy, M. (2008). *Monnaies régionales. De nouvelles voies vers une prospérité durable*. Mayers, Editions Charles Leopold.
Marra, M. (2015). The city is a bitch: The dematerialization of the Urban. *Journal of Biourbanism, 4*(1–2). https://journalofbiourbanism.org/
Marx, K. (1964). *Economic and philosophic manuscripts of 1844*. International.
Mattauch, L., Hepburn, C., & Stern, N. (2018). *Pigou pushes preferences: Decarbonisation and endogenous values* (CESifo Working Paper no. 7404).
Matthey, A. (2010). Imitation with intention and memory: An experiment. *Journal of Behavioral and Experimental Economics* (formerly The Journal of Socio-Economics), *39*(5), 585–594.
Meadows, D, et al. (2006). *I nuovi limiti dello sviluppo. La salute del pianeta nel terzo millennio*. Mondadori.
Mill, J. S. (1911). *Principles of political economy*. Longmans Green & co.
Mishan, E. J. (1967). *The costs of economic growth*. Staples Press.
Morin, E. (1999). *Les sept savoirs nécessaires à l'éducation du futur*. UNESCO.
Mulligan, M. (2002). *Ecological pioneers: A social history of Australian ecological thought and action*. Cambridge University Press.
Muraca, B., & Schmelzer, M. (2017). Sustainable degrowth: Historical roots of the search for alternatives to growth in three regions. In I. Borowy & M. Schmelzer (Eds.), *History of the future of economic growth: Historical roots of current debates on sustainable degrowth* (pp. 174–198), Routledge, London.
Navarro, V., & Torres Lo'pez, J. (2014). *Un proyecto economico para la gente*. https://web-podemos.s3.amazonaws.com/wordpress/wpcontent/uploads/2014/11/DocumentoEconomicoNavarroTorres.pdf.
Nisbet, R., & Perrin, R. G. (1977). *The social bond*. Knopf.
Paquot, T. (2007). *Elogio del lusso. Ovvero l'utilità dell'inutile*. Castelvecchi.
Pigou, A. C. (1920). *The economics of welfare*. Macmillan.
Piketty, T. (2013). *Capital in the twenty-first century*. Harvard University Press.
Polanyi, K. (2001). *The great transformation: The political and economic origins of our time* (2nd ed.). Foreword by Joseph E. Stiglitz.
Prochaska, J. O., & DiClemente, C. C. (1992). *The transtheoretical approach*. Oxford University Press.
Rappaport, J. (1977). *Community psychology: Values, research and action*. Rinehart & Winston.
Ronan, K. R., & Johnston, D. M. (2005). *Promoting community resilience in disasters: The role for schools, youth, and families*. Springer.

Sachs, W. (1999). *Planet dialectics-explorations in environment and development*. Fernwood Publishing.
Sahlins, M. (1972). *Stone age economics*. Routledge.
Sarason, S. B. (1974). *The psychological sense of community*. Jossey Bass.
Sen, A. (2011). *The idea of justice*. Harvard University Press.
Sovacool, B. K. (2014). What are we doing here? Analyzing fifteen years of energy scholarship and proposing a social science research agenda. *Energy Research & Social Science, 1*, 1–29.
Stern, N. (2006). *The economics of climate change: The stern review*. Cambridge University Press.
Turner, G. M. (2008). A comparison of the limits to growth with 30 years of reality. *Global Environmental Change, 18*(3), 397–411.
Valentinov, V., & Chatalova, L. (2013). Transaction costs, social costs and open systems: Some common threads. *System Research and Behavioral Sciences, 31*(2), 316–326.
Van Ypersele, J. P. (2015). *Une vie au coeur des turbulences climatiques*. De Boeck supérieur, Bruxelles.
Vatn, A. (1998). Input versus emission taxes: Environmental taxes in a mass balance and transaction costs perspective. *Land Economics, 74*(4), 514–525.
Vatn, A. (2005). Rationality, institutions and environmental policy. *Ecological Economics, 55*(2), 203–217.
Victor, P. A. (2014). The Kenneth E. Boulding memorial award 2014: Ecological economics: A personal journey. *Ecological Economics, 109*, 93–100.
Waller, V., Farquharson, K., & Dempsey, D. (2015). *Qualitative social research: Contemporary methods for the digital age*. Sage.
Walker, B., & Salt, D. (2006). *Resilience thinking: Sustaining ecosystems and people in a changing world* (1st ed.). Island Press.
Warner, B., & Shapiro, J. (2013). Fractured, fragmented federalism: A study in fracking regulatory policy. *Publius the Journal of Federalism, 43*(3), 474–496.
Weart, S. R. (2003). *The discovery of global warming*. Harvard University Press.
World Commission on environment and development. (1987). *Our common future*. Oxford University Press.
Yin, R. K. (2003). *Applications of case study research*. Sage.
Yin, R. K. (2009). *Case study research: Designs and methods*. Sage.

CHAPTER 4

The Application of Resilience for Human Community: From Theory to Practise

The review of the relevant literature of Degrowth theory has shown that two of the main criticisms of the degrowth position is that this paradigm lacks effectiveness and is a confusing concept. At this point the Transition Town Movement (TTM) within the paradigm proposed by Serge Latouche comes into play. According to Hopkins (2008), the TTM is based on a pragmatic approach that underlines the importance of practical activities: renewable energy facilities, local food provision, local currencies, free trade, sustainable market. These activities offer concrete outcomes for the community and encourage citizens to get involved in the process. Transition Towns frame climate change and the scarcity of fossil resources as one problem.

According to Gare (2016) and Bauman (2000), capitalism is parasitic. Parasitic organisms are always looking for a new body from which to suck the lifeblood to survive when the body to which it is attached is exhausted. However, the parasite cannot satisfy its needs without damaging the guest, thus eventually destroying the conditions of its prosperity or even its survival. According to Bauman (2007), its strength lies in the extraordinary ingenuity with which it seeks out and discovers new host species each time the previously exploited species diminish in number or become extinct. For formal economic science, exhaustion corresponds to the time when there is no new entry capital, there is stagnation of consumption, markets are saturated and thus profits reduce.

M. Carcea, *The Road Towards DemocraCity*,
https://doi.org/10.1007/978-981-16-3505-2_4

This is how Bauman explains the emergence, through credit cards, of the figure of the "eternal debtor". Starting from the promise that satisfying one's desires did not require there being sufficient money in the first instance, this tool turns the eternal debtor's life into a nightmare and enslaving the debtor with a further promise: the help of "friendly banks", which do not require the return of money lent but allow access to new forms of debt. Thus, credit, has just like parasites, thrived for a while on the financial skin of these new subjects, destroying previously non-indebted consumers now completely absorbed. The current crisis is therefore a paradigm of this parasitic system, a combination of a scrupulous financial system enabled by a complicity state, as demonstrated by the GFC, a crisis triggered by the subprime mortgage collapse, introduced on the initiative of the Clinton White House in the United States.

Resilience and Degrowth: Two Sides of the Same Coin

The Brundtland Report (1987) provides the definition of sustainable development with these words: "*Sustainable development seeks to meet the needs and aspirations of the present without compromising the ability to meet those of the future*" (WCED, 1987, p. 40). At this point, it would be useful to summarize the milestones of this new discipline:

- 1972, Limits to Growth, Club of Rome/United Nations Conference on the Human Environment (Stockholm Conference);
- 1980, IUCUN, International Union for Conservation of Nature.
- 1983, World Commission on Environment and Development (WCED).
- 1987, Our Common Future, known as the *Brundtland Report.*
- 1992, The United Nations Conference on Environment and Development (UNCED), or the *Rio de Janeiro Earth Summit.*
- 1993, Italia—Piano Nazionale per lo Sviluppo Sostenibile in attuazione dell'Agenda 21.
- 1994, The Charter of European Sustainable Cities and Towns Towards Sustainability or the *Aalborg Charter.*
- 1996, The Second European Conference on Sustainable Cities & Towns, Lisboa.

- 1997, Kyoto Protocol. An international treaty which extends the 1992 United Nations Framework Convention on Climate Change (UNFCCC).
- 1999, Italia, Conferenza di Ferrara: istituzione del Coordinamento Agende 21 locali italiane; Italia, Ministero dell'Ambiente: istituzione del Servizio per lo Sviluppo Sostenibile.
- 2000, The Hannover Conference.
- 2002, The Sixth Environment Action Programme of the European Community; Johannesburg The World Summit on Sustainable Development.
- 2009, ISO 9004, Guidance Document Managing for the sustained success of an organization. A quality management approach.
- 2010, ISO 26,000 Guidance on social responsibility.
- 2015, Paris, United Nations Climate Change Conference.
- 2019, Madrid, United Nations Framework Convention on Climate Change, COP 25.
- 2021, Glasgow, United Nations Framework Convention on Climate Change, COP 26.

Sustainable development, according to the definition provided by the Bruntland Report, implies the possibility of economic growth within certain limits. Examples of such growth include for instance, deforestation emission limits, the use of non-renewable energy sources, and so on. It can be argued that, according to the Bruntland Report, development and economic growth are basically synonyms. The conclusions of the Bruntland Report were in sharp contrast to those of the Club of Aurelio Peccei (co-founder of the *Club of Rome*) in 1972, i.e. that infinite growth is not possible on a finite planet. The Club of Rome was founded in 1968 by a group of international professionals from academic, diplomatic, and industrial fields, and from civil society who met in Rome to discuss the issues of economic growth and the indiscriminate use of resources in an increasingly globalized world.

The Club of Rome commissioned a scientific study undertaken by scientists from the Massachusetts Institute of Technology (MIT). This study, which was eventually published in 1972 under the title *The Limits to Growth*, highlights the problem of infinite growth in a finite system. Using the System Dynamics Model to understand the dynamic behaviour of complex systems, these MIT researchers created a series of possible scenarios for the future of humanity and the environment, trying to manipulate various variables. The study found that the most supported model of the behaviour of the Earth's system is overshooting and collapsing. Growth means exponential growth, i.e. the increase of a constant percentage of the whole, constant over time. In this model, according to the report, the five items that tend to increase exponentially are: population, industrialization, malnutrition, consumption of non-renewable resources, and environmental pollution. Whatever manipulation of the variables was carried out, the most optimistic estimate was that growth was projected to stop by 2100.

However, this infinite growth according to Boulding (1966), Cochet (2005), Jackson (2009), and Piketty (2013), and other scholars is destined to end. The approach currently in use has been a desperate fight against the natural limits of the planet and, most importantly, against time. Only by strengthening technology have we been able to respond to the growing pressures that natural limits impose on the global economy and the lifestyles of industrialized countries. But mankind has created an illusion. According to Hans Jonas and his "The Imperative of Responsibility: In Search of Ethics for the Technological Age" (1985), the technological age enforces an ethical change. Jonas argues that traditional moral values measure the behaviours in the relationships between human beings. According to Jonas, nowadays we have to understand that nature is one of the keystones of ethics: the consequences of reaching the limits of the exploitation of the planet will affect the poorest countries in the southern hemisphere and future generations more severely than current populations in the Global North. Technology—which had made the world inhabitable, freeing man from terror and handing them dominance of nature—threatens to overwhelm everything in its way: the main key of a project of humanization of the world, according to Jonas, has become what is dehumanizing and distorting the entire world.

The problem raised by Jonas and acknowledged by Mendelsohn (1989), are the undesirable results of technology in economic terms. Further, Jonas also links the issue of respect for the natural environment to his argument for the necessity of development ethics. Similarly, the Japanese philosopher Tomunubo Imanichi, (2009) with his "eco-ethica" explores issues arising from a changing habitat and our changing moral consciousness and presents an ethical compact that transcends interpersonal ethics so as to also encompass companies and government.

Imanichi advocates learning from nature and argues that the human race has an ethical responsibility towards nature and things, including life itself. By demonstrating that virtues were created as necessary in the past, he raises the possibility of creating new virtues to meet contemporary needs. With a firm grasp of Western philosophy and ethics, the author adds a new dimension by contributing a Japanese perspective. Every simulation and any kind of approach used by MIT started from the general assumption that population and capital are free to grow until the natural limit is reached. This is the modus operandi of the Anthropocene. Self-imposed limits and formulating viable solutions are not part of this geological era. Going back to the MIT report, it suggested that mankind has the chance to choose its own path.

Although shocking, the report in 1972 offered a margin of time in order to find a solution. Today, solutions seem to be taken from science fiction novels as birth control or the colonization of other planets, a pioneer in this field was the utopian novel "Red Star" of Alexander Bogdanov (1980). Such theories are becoming more and more concrete. All this because man refuses to acknowledge the impossibility of infinite growth in a finite planet. However, if man does not learn to recognize the limit, nature will do it. This then is perhaps the junction of the long-standing problem: opportunity, crisis, and transition. Crisis: in Greek (krino) means separate, i.e. choose the good part not the bad part. Crisis is a double moment by definition, it indicates a transition: we are the ones who decide which one is the good part and which is bad.

The report made by the Club of Peccei itself underlined the word "transition", i.e. moving from a state of imbalance, framed as growth, to a one of balance. What the report wanted to underline, and it is still the subject of criticism for the advocates of Degrowth, that Degrowth does not mean economic stagnation, but a state of balancing forces of the opposite direction. No more waste of energy for unlimited production but an understanding of the delicate relationship between human and natural limits. The Club of Rome has not designed the city of tomorrow, in a sense MIT analyst accepted Ghandi's lesson: Earth has enough for everyone's needs, but not for the greed of some people. We can survive as a species, according to Vandana Shiva (2000), only if we live according to the laws of the biosphere. The biosphere can satisfy everyone's needs if the global economy respects the limits imposed by sustainability and justice.

The Detoxification of the Imaginary

According to Frans de Waal (2006), this is precisely the challenge of Degrowth: understanding not whether or not we will be able to manage overpopulation, but if we are able to share resources with honesty and fairness. This is the very goal of the Transition Town Movement as well and is also its undeniable bond with Degrowth theory: promoting awareness regarding global environmental, social and economic issues, supporting local action from below (bottom-up) and stimulating the creativity of local communities to choose a path utilizing knowledge (degrowth), tools (transition) and resources (resilience) in order to reach a realistic, long-term achievement that can lead mankind to a finally balanced society.

The difference between two such studies is best extrapolated by the *Bruntland Report*:

> Far from requiring the cessation of economic growth, it recognizes that the problems of poverty and underdevelopment cannot be solved unless we have a new era of growth… To bring developing countries' energy use up to industrialized country levels by the year 2025 would require increasing present global energy use by a factor of five. The planetary ecosystem could not stand this, especially if the increases were based on non-renewable fossil fuels. Threats of global warming and acidification of the environment most probably rule out even a doubling of energy use bared on present mixes of primary sources… Any new era of economic growth must, therefore, be less energy intensive than growth in the past… (WCED 1987, Ch.1)

The Degrowth theory, the daughter so to speak of the Club of Rome, starts from this very paradox generated by the *Bruntland Report*: there is no such thing such as an economic growth without massive depletion of the natural resources of the planet. As Serge Latouche argued (2011) development cannot be sustainable, unless we want to create an oxymoron. Paraphrasing the French economist, economic growth is not only unsustainable, it is not even desirable. The future lies in frugality and solidarity. We have already faced the definition of Degrowth, and the anthropocentric "mistake" as proposed by Bruntland Report and its definition of sustainability.

The transition towards a society based on degrowth arises from this question: What right does mankind have to exploit nature? The role of ethics in the theoretical framework, the basis of that concrete utopia synthesized in the realization of *DemocraCity* is a tool, or even better, an inner value that has its roots in Aristotelian and Plato's philosophical visions. Since the *koinonia*, ethics is a necessary tool in order to realize what in economics is called profit.

In the previous chapter, we compared the vision of Plato and Aristotle. However, we believe that the moral question raised by father of the *lassez-faire*, Adam Smith, must also be addressed. The pioneer of liberalism took the lesson of Aristotle; Smith's moral philosophy is based on the clear distinction between selfishness and self-interest: the first is the achievement of personal profit at any cost, even abusing others in the quest for profit; the second one is the achievement of the satisfaction of needs without harming others. According to Smith in his *An Inquiry into the Nature and Causes of the Wealth of Nations* (1776) and his *The Theory of Moral Sentiments* (1759) the complexity of the economic system is the result of human complexity consisting of sentiments, moral and motivational principles.

The complexity of self-interest, if structured by moral limits and supported by the diversity of motivational choices, could create an imaginary balance that coincides with the collective interest. "The invisible hand" does not depend only on factors within the economy, but also on those of a complex system of moral and political limits. If these limits are lacking, individual interest degenerates into selfishness and the system will collapse. Only by respecting an external limit and an internal limit, is it possible to achieve a balance between self-interest and collective interest. It is the rediscovery of this limit, identified in the "instrument" of resilience, which is the basis of this research path. The economy cannot prevaricate either on man or natural laws.

According to Gadrey and Jany-Catrice (2005), a new social and economic system is needed for the purpose of taking care of the community, creating more durable goods, improving the stages of repair and recycling. A new way of thinking about nature, social cohesion, and the quality of work and of democracy is required. Taking care of people, communities and goods takes time. Though this might be considered "wasted time" from the point of view of "productivity", this "time" is needed to achieve a sustainable society (Gadrey & Jany-Catrice 2006). Latouche et al. (2001) agree with another key element of the concept of resilient communities, the concepts of social and human capital. Human capital can be identified within people, groups, networks, voluntary organizations, and services in the community.

Global capitalism according to Bauman (2009), does not feel bound by any ethical or institutional limits. Human organizations based on this model, according to Gare (2017) and Korten (2000) have become cancerous tumours in the global ecosystem. We desperately need a new ethical order to move from a world of dependent consumers to one of the responsible producers, transforming ourselves into healthy cells that are living in a tumour. "As healthy cells can bring about spontaneous remissions, so, hopefully, can such inspired people. To do so requires engagement in politics" (Gare, 2017, p. 146).

Paraphrasing Bauman (2009), we can choose to oppose the imbalances and dependencies that are forced on global citizens, even if we cannot oppose globalization as a phenomenon that dominates individual choices. Economic and political choices released from the awareness of ecological limits and the choice of ethical principles do not guarantee human survival in the long term, rather they condemn it to submit to the future limits that will be imposed by nature itself (Vatn, 2005).

Ready and Resilient: Resilience, Degrowth, and a New Concept of Sustainability

The term resilience comes from the Latin *resilire* meaning bounce back. The term usually indicates the ability of a substance or object to spring back into shape and the concept of resilience has been redesigned by social sciences. It combines the capacity to withstand rapid and violent change, the implementation of a strategy, and adaptation to it. Resilience and adaptability are the key elements for an open and dynamic system, a system subjected to sudden changes. In the specific case of human communities, according to Mulligan (2002), resilience is characterized by the capacity to react in difficult situations. Communities are oriented towards the common good and guided by principles of solidarity, cooperation and collaboration. These principles help to develop mutual aid, mobilization of relational resources, promotion of ideas such as sharing responsibility to create welfare for people living in the same community (territorial, social, and so on).

When communities are subjected to critical situations, the consequences are often considered exclusively as negative. In these situations, the general opinion is that communities are unable to handle a crisis without external assistance (Sonn & Fisher, 1998). According to Lavanco and Novara (2012), however, the previous statement is a devaluation of potentialities and internal resources within communities. Some scholars identify resilient communities based on specific circumstances. For instance, Gist and Lubin (1989) provide a different idea of resilient communities. They consider as resilient communities only those affected by dramatic events such as disasters. Tobin and Whiteford (2002) highlight the "therapeutic role" of the crisis within the community; the crisis is considered as a catalyst for solidarity, social cohesion and sense of community among members. The "therapeutic role" has a short-term effect, a precise reaction of the community against a critical event. The significance and the importance of resilience does not only have great impact in social studies, it has also influenced environmental and ecological fields of study. According to Norris et al. (2008), resilience is a process that creates a network of adaptive capacities. For instance, there are resources with dynamic attributes such as Robustness, Redundancy, and Rapidity (3Rs).

According to Kimhi and Shamai (2004), resilient communities have three main characteristics: (1) Propensity for resistance (the ability of a community to absorb an impact); (2) Recovery trend (indicates the speed and the ability to recover from stress); (3) Propensity for creativity (the creative potential of social systems to improve their performance as a result of adversity). Kendra and Wachtendorf (2003) argue that resilience of communities includes not only what happens after a critical event, but also the importance of community preparedness. According to the model of community resilience by Ronan and Johnston (2005), called SS4R Systems Strengthening (Risk Reduction, Readiness, Response, Recovery) Prevention Model, a key factor is the prevention of disasters by strengthening systems used for risk reduction, readiness, response, and recovery. The model of Ronan and Johnston explains not only what happens after a disaster in resilient communities, but also, the role of the community preparedness for such a negative event.

Within the dedicated literature on resilience of communities, according to Prati and Pietrantoni (2009), there are some concepts that overlap, such as community competence and empowerment. Sonn and Fisher (1998) indicate that the extension of the concept of resilience from the individual to the community is represented by the research area on competent communities. Resilience and competence are two different concepts, even if they partially overlap. The first one refers to the process of adaptation following a risk factor or a disturbing event, while the latter focuses more on coping skills and community resources to deal effectively with adversity. The concept of competence within a community focuses on the characteristics of a given community, similar to the first conceptions of resilience as a personality trait.

Resilience of community, instead, denotes a process of adaptation that, from an ecological point of view, does not take into account factors within the community, but also those external to the community itself, such as international aid or relations with the neighbouring social and political entities. For instance, Prati and Pietrantoni (2009) argue that a policy of aid implemented in a manner that is not respectful of the context and of the recipients can be a further risk factor that neutralizes potential elements of resilience inherent in a community; in the study of resilience internal and external factors within the community must be considered.

Moreover, even empowerment and community resilience may have overlapping areas and there may be the possibility of integration. According to Zimmerman (2004), resilience and empowerment share the emphasis on factors such as participation, mastery, and involvement. However, individual resilience focuses on adapting to risk factors, while empowerment focuses on social justice and the concept of a competent community. According to Norris et al. (2008), community resilience can be framed as a set of resources whose dynamic characteristics are to be understood in terms of robustness, redundancy, and rapidity: robustness refers to the property of the element to resist stress without deteriorating. Redundancy refers to the ability of an element to be replaced with another in case it deteriorates due to stress. According to the authors, redundancy also includes the diversity of resources that the community relies on.

The last attribute is rapidity, which is the speed in accessing and using resources. The literature indicates other concepts relating to the elements that make up the resilience of communities. Breton (2001) for instance, refers to social and human capital, i.e. people, groups, networks, voluntary associations and services in the community. Sonn and Fisher (1998) emphasize the importance of settings such as churches, networks of extended families, sports associations, and recreational groups. Another model that protects the integrity of the community is the one proposed by Bachrach and Zautra (1985). This model includes three variables: self-efficiency, problem-oriented coping, and a sense of community as predictors of greater involvement in solving a community problem. The sense of community is an important variable because it can give meaning to the collective action to be taken to overcome the problem effectively.

Other Ways and Other Worlds

Our current society dominated by mercantilist logic is characterized by a very low level of resilience. Paraphrasing Latouche (2006), we all live dependent on the evil carousel. Being resilient means being connected to the whole world from an ethical point of view. Tools of resilience are the collaborative spirit and the inclination to share, not the dependence. Resilience is a cornerstone concept of Transition as opposed to the concept of dependence. As mentioned, the transition movement takes inspiration from the psychology of community. Transition offers a positive vision of the future; its aim is creating a better and a practical alternative. Being resilient nowadays and in the future means having a set of social, economic (and physical) resources within a community that can meet most demand independently.

As well as for a society based on Latouche's degrowth, transition does not aim to create retrograde communities or a return to the stone age. It wants to create cities that are able to withstand the shocks due to such network system: multiply renewable energy sources, reintroducing vegetable gardens, polyculture, forestry, and proximity farming through the purchase of common land to support the start of activities by young farmers; the spread co-ownership of gardens in the suburbs of urban centres with the organization by farmers of shared selling facilities. These initiatives allow the establishment of a local, social and supportive economy, favouring the maintenance or reintroduction of agricultural employment.

According to Kempf (2010), such actions will assist in reconstructing spheres of autonomy, in which individuals, families, and communities can satisfy part of their needs without having to resort to the market. The exchanges will decrease and with them the pollution caused by the transport of goods; this will also improve the environment, in fact people will generally take care of the resources on which they depend. In addition, people will regain creative control of their lives and will reduce the neurotic frustration that characterizes capitalism as being endangered of becoming extinct. People need to be the starting point of any transformation because people acting together can transform what seems immutable.

It is possible to escape from capitalism, Latouche argues (2006), only through the recovery of collective practices and the construction of new alliances in the self-organization of growing parts of companies subjected to the mechanisms of accumulation and valorization of capital. It is a social body under construction in which class composition has been profoundly modified. The attack on labour in the large concentrations of workers today follows the attack on public employment: old and new workers, workers of services and knowledge, of material or immaterial work intersect with the great migratory processes.

The market economy, according to Bertorello and Corradi (2011) eats the so-called workers' aristocracies until they reach important segments of the middle classes, giving rise to social remixing that creates new conditions of proximity and mutual recognition, from the potentially destabilizing value of the system itself. We need to start from a community that can also counterbalance internal shocks, for instance what Latouche (2007), calls a momentary degrowth of the productive sector, since the territory is not mono-productive and the skills of the people who live there are not mono-sectoral.

According to Latouche (2007), productive specialization is a serious economic damage on a global scale. Although according to Adam Smith's logic, it allows us to produce more quantities by reducing fixed costs, it is extremely vulnerable in case of saturation of the market and production stops. However, the challenge posed by degrowth and transition is also this: to reconvert infrastructures. Moreover, it is more than possible that, initially, a policy based on degrowth may, paradoxically, lead to an increase in production, due to a demand for ecological products and machinery. An economy based on renewable energies would involve the development of productive sectors such as the construction of wind and related turbines, photovoltaic cells, hydrogen engines, bicycles, and light metros in addition to a widespread increase in organic agriculture and reforestation.

A Different Globalization

According to Latouche, the reductions, re-use, repair and recycling that would result in the abandonment of planned obsolescence would also create new activities whose expansion would give rise to a positive balance of employment, with the emergence and diffusion of new jobs, from forest expert to eco-architect. Marx had already foreseen the limits of hyper specialization of human labour in 1884. In his *Economic and Philosophic Manuscripts of 1844* (1932), human work instead of being a positive realization of the spirit and of human energy, is estrangement and alienation (*Entfremdung*) both during the productive activity and in the relationship with the products of labour.

In this sense, hyper specialization, favouring a reductionist approach based on the splitting of knowledge and reality, would instil in men the conviction that, once the activities related to their specific professional role were carried out, they would be exempt from all political duties and social rights that would compete as citizens, towards other members of the community and their state. It is in this context that the reflection of Edgar Morin's "Seven Complex Lessons in Education for the Future (1999), not only provides an essential contribution to the understanding of the limits and risks, at the political and social level, of an education marked by hyper specialization, but represents an important attempt to outline appropriate responses to the multidimensional character and not divisible from the problems of contemporaneity.

According to Morin (1999), there is a real growing democratic deficit due to the appropriation by experts, specialists, and technicians, of an increasing number of vital problems. Knowledge has become more and more esoteric (accessible only to specialists) and anonymous (quantitative and formalized). In this way, the expert becomes progressively more and more incapable of assuming a multidimensional perspective and the citizen loses the right to knowledge able to contextualize, connect and integrate. Starting from premises similar to those of Morin, and in particular from the observation of the irreducible complexity of reality and the critique of the hyper specialized fragmentation of knowledge, Howard Gardner (2011), using a multidisciplinary perspective, identifies five types of intelligence (mind) that individuals should develop to be able to adequately face the challenges of the future in an interconnected, technological and globalized world.

The first three types of intelligence have to do with the forms of knowing and are: the *disciplined mind*, in the double meaning of mastery of a discipline and ability to renew the latter through a regular and rigorous application; the *synthesizing mind*, since, in the face of the uninterrupted and frenetic multiplication of information and the mass of knowledge, it is necessary, for individuals as citizens, to be able to perform synthesis processes; the *creating mind* that underlies innovation.

The last two types of intelligence concern the sphere of interpersonal relationships and are the *respectful mind* that rejects intolerant attitudes to recognize and welcome the differences that exist between individuals and different societies; the *ethical mind*, that is the conscious assumption of responsibilities related to one's multiple roles (professional, citizen, member of a family). These intelligences (minds) are not specific abilities but they are complex and multi-faceted in that they govern not only our cognitive processes, but also the moral dimension and therefore human actions and behaviours, ending up as assuming a political value and meaning.

Another Global Space

Gardner (1991) believes that they must become the primary aim of a lifelong learning process that is not exclusively entrusted to traditional educational structures—the school—but extended to companies and to the liberal professions. The reflections of Morin and Gardner, in their similarities and differences, clearly show how the identification of adequate and effective answers to the problems posed by the complexity and irreducibility of the world in which we live, represents an urgent need: survival of democracy and democratic values.

If hyper specialization leads to a reduction in the sense of responsibility of citizens and inhibits social solidarity, it is essential to establish a pedagogical solution—with clear political values—that, through a reform of thought and teaching, creates responsible citizens, supportive and capable of dealing positively with complexity, in all directions in which it is currently realized and will develop in the future, avoiding reductionist approaches, destined to prove to be unsuccessful. Resilience is applicable in many areas and is a fundamental tool: for the full realization of man in the humanistic sense, for the full success of the exchange in the economic sense, for the survival of the species in the biological sense.

However, resilience is above all awareness of a limit. The concept of resilience is the opposite of dependence. Dependence knows no limits. It can be said that our society knows no limits and therefore it is not a resilient society. The transtheoretical model of behaviour change developed by Prochaska and DiClemente (1982), also known as the "Stages of Change Model" in the late 1970s focuses on the decision-making of the individual and is a model of intentional change. This model operates on the assumption that people do not change behaviours quickly and decisively. Rather, change in behaviour, especially habitual behaviour, occurs continuously through a cyclical process.

The transtheoretical model is not a theory but a model; different behavioural theories and constructs can be applied to various stages of the model where they may be most effective. In conclusion, facing a problem that requires change, the implementation or otherwise of change depends on understanding and overcoming resistance to change.

Similar to Georgescu-Roegen's approach which focuses on the laws of thermodynamics and entropy, Kurt Lewin and Cartwright (1951) utilizes the third law of dynamics, that is every action or force always elicits an equal and opposite reaction, in his theoretical approach. According to Lewin and Cartwright (1951), a group can be framed as a dynamic force field, in which equilibrium is produced by equal and opposite forces. Lewin argues that the dynamics in the force field can be changed by adding forces in a particular direction or by decreasing the opposite forces.

Again, according to the Lewin, a level of equilibrium is achievable in both cases, however, the choice between the two alternatives implies different effects: high tension in the social, economic, environmental field in the first case, low tension in the second one. There are two ways to overcome resistance: increase the pressure in the opposite direction to the resistance or decrease the resistance. The second alternative is the one preferred within the transtheoretical model of behaviour change framework. Motivational interviewing is an interview methodology that has been developed by W. R. Miller and S. Rollnick in the late 1980s and aimed primarily at people with pathological dependence.

According to Hopkins (2008), it is a conversation aimed at reinforcing the motivation of a person's commitment to change. The transition is therefore based on the dependence that the current society has towards oil. It proposes to overcome this stage through paths and processes similar to those used for toxic substances. The first step towards the changing of our collective imaginary is therefore the strengthening of resilience.

Permaculture: The Basis of a Resilient Community

There is as we have seen, copious literature on EE; and resilience seems to follow the same path: both are terms so abused as to have lost their meaning. In this chapter, following the analysis of degrowth, we have re-introduced the topic of TTM, and we have dealt with the controversial term of resilience. However, as degrowth finds its concretion in resilience, it finds its materialization in permaculture. According to Mollison and Holmgren (1978), permaculture is not a method of organic gardening or sustainable agriculture, energy saving building or environmentally friendly development. Permaculture is an ethical science for the long-term design of the natural and social environment. This vision emphasizes the seminal part of the ecosystem design for a balanced interaction of species and biodiversity within them and is based on values, perspectives, projects, and holistic management approaches. This "new" ecological conscience, as we already analysed during the review of the literature, was introduced in 1972 with *Limits to Growth*, but also with the oil crisis of 1973 and 1979. The Permaculture concept was created by two Australian scholars, David Holmgren and Bill Mollison who published *Permaculture One* in 1978. The term coined emphasizes how such a concept embraces a more holistic (and human) achievement (culture) rather than a concept mutated from an agricultural and environmental approach (agri-culture).

The *crasis* permaculture suggests a permanent culture or a knowledge to last over time. Despite the terms resilient community, resilience, permaculture, etc. Being—erroneously and not, according to Robinson & Carson (2015)—a phenomenon of recent interest (academic, political, social and so on) permaculture has its roots in the work of Masanobu Fukuoka. The ethical and ecological cornerstones identified by the Japanese author share common ground with permaculture. Through forty years of studies of the environment and nature, Fukuoka as a farmer and natural philosopher came to the conclusion of doing nothing, i.e. to minimize the human footprint on nature (*The One-Straw Revolution: An Introduction to Natural Farming*, 1983).

Permaculture is a countermeasure (from the bottom) to counteract human interference with nature on a global scale. For this reason, permaculture is one of the first solutions found by resilient communities who had adopted the R of Relocalize, contained in the degrowth paradigm. Relocalizing agriculture is one of the bases of permaculture. According to Holmgren (2010), the current industrialized society has fully invested in the agriculture sector, transforming it into the most fuel-dependant production sector. Resilience and permaculture have to be included in the broader paradigm of degrowth because they are fundamental keys to the transition from well-being and wellness (I like to refer to such matter as *Well-Being/Well-Having*). The path of *homo resaliens* must be guided by a holistic and transdisciplinary discipline that combines natural sciences, economics, politics, psychology and philosophy. It is one that can overcome the constructive paradigms of individual disciplines, unsuitable to explain complex phenomena. This discipline has very deep ethical roots: it is based on the principle of responsibility for the

Earth and future generations, on reciprocity and social solidarity. Fundamental therefore, is the lesson of A. Bogdanov contained in his *Tektology*, the basis of the *System Theory and Universal Science of Structures* (The Russian Marxist has also provided a widest definition of crisis). According to the Russian philosopher, the organization—performed by each single system for itself—previously specified only into individual living organisms and in their hierarchy and classification, basically it represents the human being and the whole of nature.

Nature represents the prime engine and according to Bogdanov, nature is part of every single human process (from growing complexity to the social experience) but, in addition, nature is within those entities considered elementary in the past (Gare, 2000). Thus, it seems quite clear that permaculture's ethic is not framed in an anthropocentric view, it does not consider mankind superior to other species. Perhaps permaculture considers mankind as a privileged inhabitant of Gaia, and for this reason it involves a principle of superior responsibility. Mollison and Holmgren (1978), pioneer of the subject, was able to synthesize the twelve principles that define the permaculture ethical framework:

- Observe and Interact: beauty is in the mind of the beholder.
- By taking the time to engage with nature we can design solutions that suit our particular situation.
- Catch and Store Energy: make hay while the sun shines. By developing systems that collect resources when they are abundant, we can use them in times of need.
- Obtain a yield: you can't work on an empty stomach.
- Make sure that you are getting truly useful rewards as part of the work you are doing.
- Apply Self-Regulation and Accept Feedback: the sins of the fathers are visited on the children of the seventh generation.
- We need to discourage inappropriate activity to ensure that systems can continue to function well. Negative feedback is often slow to emerge.
- Use and Value Renewable Resources and Services: let nature take its course.
- Make the best use of nature's abundance to reduce our consumptive behaviour and dependence on non-renewable resources.
- Produce No Waste: waste not, want not or a stitch in time saves nine.

- By valuing and making use of all the resources that are available to us, nothing goes to waste.
- Design from Patterns to Details: can't see the forest for the trees.
- By stepping back, we can observe patterns in nature and society. These can form the backbone of our designs, with the details filled in as we go.
- Integrate Rather Than Segregate: many hands make light work.
- By putting the right things in the right place, relationships develop between those things and they work together to support each other.

- Use Small and Slow Solutions: slow and steady wins the race or the bigger they are, the harder they fall.
- Small and slow systems are easier to maintain than big ones, making better use of local resources and produce more sustainable outcomes.
- Use and Value Diversity: don't put all your eggs in one basket. Diversity reduces vulnerability to a variety of threats and takes advantage of the unique nature of the environment in which it resides.
- Use Edges and Value the Marginal: don't think you are on the right track just because it's a well-beaten path.

  The interface between things is where the most interesting events take place. These are often the most valuable, diverse, and productive elements in the system.
- Creatively Use and Respond to Change: vision is not seeing things as they are but as they will be.

We can have a positive impact on inevitable change by carefully observing and then intervening at the right time. We have already discussed that the well-having does not coincide with the well-being of the individual. According to Latouche (2007), despite the increase in per capita consumption, real progress in wealth has diminished, because an even larger share of the wealth produced must be used to counteract the damage caused by overdevelopment, for example in areas such as human health, environment and so on. Economic growth requires more energy to produce and even more energy to counteract the damage of production itself. To create an antidote, we have to create a disease.

The following reference frameworks of permaculture can be identified:

- Georgecu-Roegen's Bioeconomy.
- Latouche's Degrowth theory.

- Hopkins' project known as Transition Towns.
- Mauss' Gift theory.
- The paradigm proposed by Kimhi and Shamai (2004) in which the two scholars propose three main characteristics of resilient communities: (1) Propensity for resistance (the ability of a community to absorb an impact); (2) Recovery trend (indicates the speed and the ability to recover from stress); (3) Propensity for creativity (the creative potential of social systems to improve their performance as a result of adversity. Permaculture can be defined as a system of planning for eco-sustainable human settlements, based on the centrality of agriculture and on particular attention to the territory. It can also be defined as applied ecology, whose reference principles are extrapolated from the observation of nature. Transition is based on the general principles of permaculture, however permaculture is based on the foundation principles of degrowth, although the key to read is implicit.

What has emerged from the research is that (key) concepts such as transition, resilience and permaculture are positive, and it is very easy start a conversation with the members of Transition Darebin. The aim of the TTM is to create resilient communities, relocate the resources of the community, develop self-sufficiency at the community level, and choose energy degrowth as a desirable future.

The Importance of Communities in a Resilient Society

Resilience is not only a tool to understand the limit framed by degrowth, it is also the ability of a substance or object to spring back into shape. The concept of resilience has been redesigned by social sciences. It combines the capacity to withstand rapid and violent change, the implementation of a strategy and adaptation to it. Resilience and adaptability are the key elements for an open and dynamic system, a system subjected to sudden changes. However, resilience is also the ability to learn from the experience. In 1974, the psychologist Seymour Sarason introduced the concept of psychological sense of community. He proposed that it become the conceptual centre for the psychology of community, stressing that psychological sense of community "is one of the major bases for self-definition" (p. 157).

Other studies have followed the work of Sarason, and among them, the theories of Psychological Sense of Community, proposed by McMillan and Chavis's (1986). It is by far the most influential and is the starting point for most of the recent research on psychological sense of community. In order to build a psychological sense of community, McMillan & Chavis (1986) decided to synthesize with "Sense of Community". The two scholars chose and provide the following one-sentence definition: "*Sense of Community is a feeling that members have of belonging, a feeling that members matter to one another and to the group, and a shared faith that members' needs will be met through their commitment to be together* " (1986, p. 9).

McMillan and Chavis (1986) propose that a sense of community is composed of four elements: *membership, influence, Integration and fulfillment of needs and shared emotional connection*. The first value of sense of community is the importance of the membership in a community. McMillan and Chavis (1986) identified five attributes:

1. *Boundaries*
2. *Emotional Safety*
3. A Sense of Belonging and Identification
4. *Personal Investment*
5. *A Common Symbol System*

Boundaries refer to items such as language, dress, and ritual, indicating who belongs and who does not. The individual must be able to identify where the community begins and ends, or who is part of it and who is not. These boundaries can be physical, the belonging to a territory for example, but they can also be cultural, the groups are differentiated through the use of symbolic systems shared by the members. Especially in groups that have boundaries that are less than obvious, deviants or outsiders may be held in lower regard or even denounced or punished. Beyond the recognition of the boundaries, there must be a conscious identification in the group and a feeling of being accepted by the other members. McMillian and Chavis (1986) acknowledge that boundaries is one of the most complex features of the membership definition but point out that "*While much sympathetic interest in and research on the deviant have been generated, group members' legitimate needs for boundaries to protect their intimate social connections have often been overlooked*" (p. 9). The other four attributes of membership are *emotional safety* (also known as security i.e. the will to explain how a member really feels), *a sense of belonging and identification* (expectation or faith that a person will belong, will be accepted by the community), *personal investment* (increased by the personal commitment with which individuals invest in the community).

According to McMillan and Chavis, this happens because of the need for coherence between cognition and behaviours such as the greater the commitment, the more you feel part of the community; this is similar to the fact that in groups where it is more difficult to enter, you then experience a sense of higher belonging), and *a common symbol system.* According to Nisbet and Perrin (1977) understanding *a common symbol system* is a fundamental prerequisite in order to understand the community. Paraphrasing the two authors, a symbol is for the social world what a cell is for the biotic environment and an atom to the physical one "*The symbol is the beginning of the social world as we know it*" (Nisbet & Perrin, 1977, p. 47).

Influence, according to McMillian and Chavis (1986) is a multidimensional concept. Members of a group must feel empowered to have influence over what a group does, and group cohesiveness depends upon the group having influence over its members. It is essential that neither direction of influence prevails as it is the balance between conformism and individual freedom that allows the community to remain united without suffocating individual members through excessive control and pressure for equality.

Integration and fulfilment of needs, McMillan and Chavis (1986) argued that members of a community are motivated to feel part of communities that meet their needs, both practical and psychological (members of groups are rewarded in various ways for their participation). The satisfaction of the needs of the individual requires the integration of individual needs. It follows an inevitable interconnection between the members of a community. According to Rappaport (1977), a rich community, in terms of resources and status, easily arouses a stronger and lasting sense of belonging.

Shared emotional connection represents the emotional bond that unites the members and binds them to the group. This link is mainly based on interactions between people, their number and their quality: the greater the number of interactions and the better their quality, the more easily individuals will develop a community emotional bond. Furthermore, McMillan and Chavis (1986) have listed seven important features of shared emotional connection, citing relevant research for each of them:

1. *Contact hypothesis*: A deeper personal interaction increases the likelihood that people will become close.
2. *Quality of interaction.*
3. *Closure to events*: Doubtful relationships and unresolved tasks inhibit group cohesiveness.
4. *Shared valent event hypothesis*: Increased importance of a shared event (for instance a crisis) facilitates the creation of a bond.
5. *Investment:* A community become more important to someone who has given more time and energy to it thanks to a boundary maintenance.
6. *Effect of honour and humiliation on community members*: Someone who has achieved something and was awarded in front of the entire community feels more attracted and attached to it, however, for the same dynamics, someone who has been humiliated feels less attraction to the community.
7. *Spiritual bond*: McMillian and Chavis (1986) argued that this quality is difficult to explain, however maintain that it is "*present to some degree in all communities*" (p. 14), and give the example of the concept of "*soul*" in the formation of a national black community in the U.S.

Ten years later, in 1996, McMillian proposed a reinterpretation of his previous model, however, it does not substantially change the structure of the sense of community.

After the process of rereading, McMillian elected to change the names of the four dimensions, stressing the meaning that these take on a personal level and emphasizing the cultural aspects.

Membership becomes *Spirit*.
Influence is replaced by *Trust*.

Integration and fulfilment of needs is renamed *Trade* (the mutual exchange of services framed with its own practical and symbolic economy);
Shared the emotional connection is replaced by *Arts*.

Spirit

Spirit has four categories that trigger our mind:

1. *Boundaries*
2. *Emotional safety.*
3. *Sense of belonging.*
4. *Personal investment/dues paying to belong.*

Belonging to a community (*membership*) allows a subject to establish boundaries, allowing such to acquire and strengthen emotional safety. But that is not all, sub mechanisms are triggered as well: such member develops a connection to others in which he is protected from shame, opens himself (*self-disclosure*) and in addition, he can express himself as he pleases (*telling the truth*). A member influences the community through *sacrifice*, for example by dedicating time, energy, and expertise (*dues paying*). Vice versa, the community influences members through acknowledgments (or *entitlements*) and *acceptance*. Acceptance generates *attraction* and *attachment* of that person (*sense of belonging*) to the community. Just as Latouche speaks of the virtuous circle of *8Rs*, McMillian (1996) argued that the spirit of community is triggered by a virtuous mechanism between a member and his community: the acceptance of a member by the community increases his attachment and commitment to it; on the other hand, a failure means a possible humiliation and will bring with it the risk that a subject feel detached from the community.

Trust

Being part of a community is like lighting a fire: the spark is fundamental, however, keeping the fire alive is the most important part. *Trust*, according to McMillian (1996) is the tool that allows keeping such a fire alive, trust is the main engine of the sphere of influence. Once trust is created between members and community, it is necessary to find the balance between *authority structure* and the *allocation of power*. The author argues that the desirable representation of power is the cooperation in which values such as fairness and equality would create a democratic authority based on social justice. According to Coluccia (), communication, intended as common action (cooperation) is a cornerstone in order to create a greater cohesion among the members of a community. This value combined with the propensity for creativity as set out by Kimhi and Shamai (2004) would create a virtuous circle in which justice, trust, order, and authority will maintain the spark of cohesion.

Trade

According to McMillian (1996) and Dacrema (2010) the most revolutionary and far-reaching social invention for mankind was, that the advent of exchange. According to Plato, who raised the question regarding the most revolutionary invention in his *Republic*, similarities are not enough in order to create the development of a community. A community needs an exchange system, trading with others for material goods, services, or immaterial goods such as knowledge and time.

Arts

McMillian (1996) argued that the aforementioned three dimensions create a shared cultural heritage, strengthened by the existence of spiritual bonds such as sharing a religious faith or belonging to the same ethnic group. The emotional connection is replaced by the *arts*. According to McMillan, the immaterial emotional connection is concretized by sharing symbols, rituals, stories, and social representations, but also a historical memory, which helps to maintain and pass on the emotional bond between members.

What Is the Transition? Utopia is Real

Starting from van den Bergh's (2011) critique, analysed in the previous chapter, and taking inspiration from the paper *"In defence of degrowth"* (Kallis, 2011), the aim of this paragraph is the concretization of the real utopia. Mankind, one of the many species until the last century, has become a geological agent capable of transforming the structure of the biosphere: the concentration of carbon dioxide in the atmosphere is the highest it has ever been for about a million years, more than 40,000 artificial dams heavily modify natural sedimentation, the acidity of the oceans is increasing rapidly while the disappearance of species is comparable to that which destroyed the dinosaurs at the end of the Cretaceous period. Hervé Kempf argued that we need to start a new era, a new society, unless we prefer the fate of the dinosaurs.

Degrowth theory has its roots in the academic field and develops as an ideological challenge to the dominant economy, i.e. ultraliberalism. The slogan of "degrowth" appeared as a "semantic bomb" or an "explosive term" quoting Paul Aries, capable of destroying the weak consensus of submission to the dominant productivist order, or in other words of initiating a decolonization of the imaginary. To paraphrase Castoriadis we need to create an imaginary which has an unparalleled importance in the past, a creation that would contain more meaningful goals than the expansion of production and consumption, goals recognized by human beings as worth of living for (1996).

This is the immense difficulty we are faced with. We should want a society in which economic values have ceased to be central (or unique), where the economy is put back into its place as a mere means of human life and not as its ultimate end, a mad rush towards the indefinite growth of consumption. All this is not only necessary to avoid the definitive destruction of the terrestrial environment, but also and above all to free ourselves from the psychological and moral misery of our contemporaries (Castoriadis, 1996).

In conclusion, the paradigms of degrowth and transition can be considered complementary. Both contribute, with different approaches, strategies, and on different levels, to the promotion of a radical change of our collective imaginary towards more frugal and resilient communities, and lifestyles which feature solidarity. Degrowth theory opposes resistance, questioning the limits of the economic system based on unbridled capitalism, transition instead wants to stem the pressure of economic growth by creating local communities, strengthening the resilience of the economic, social and ecological elements. By linking degrowth to the theme of limit, impossibility of infinite growth on a finite planet can be demonstrated. Transition, on the other hand, encourages citizens to implement actions that respect these limits, based on the ethical and design principles of permaculture.

The aim of degrowth is, to deconstruct the capitalist imagination and to create a new one, perhaps as Crouch (2000) argues aimed at a social revolution. The transition movement follows *La force tranquille* of Leon Blum (1936). Degrowth is framed as a bastion of defence and intellectual resistance against the attack of ultra-capitalism and globalization; the transition instead aims to create resilient communities which identify the limits of capitalism and respond by reducing consumption, promoting local autonomy and self-efficiency. Paraphrasing Holmgren (2010), self-efficiency is an invisible boycott tool that slowly erodes the market and its collective imaginary based on the domination of nature which creates what Latouche (2004) calls *the diabolical carousel*, i.e. dependence on the system itself.

References

Bachrach, K. M., & Zautra, A. J. (1985). Coping with a community stressor: The threat of a hazardous waste facility. *Journal of Health and Social Behavior, 26*, 127–141.

Bardi, U., & Lavacchi, A. (2009). A simple interpretation of Hubbert's model of resource exploitation. *Energies, 2*(3), 646–661.

Bauman, Z. (2000). *Liquid modernity*. Cambridge.

Bauman, Z. (2007). *Consuming life*. Cambridge.

Bertorello, M., & Corradi, D. (2011). *Capitalismo tossico. Crisi della competizione e modelli alternativi*. Edizioni Alegre.

Bogdanov, A. (1980). *Essays in tektology: The general science of organization* (G. Gorelik, Trans. [1980]). Intersystems Publications, Seaside.

Boulding, K. (1966). The economics of knowledge and the knowledge of economics. *American Economic Review, 56*(1/2), 1–13.

Breton, M. (2001). Neighborhood resiliency. *IEEE Journal of Community Practice, 19*(1), 48–61.

Castoriadis, C. (1996). *La montée de l'insignifiance* (p. 96). Les carrefours du labyrinthe IV.

Cochet, Y. (2005). *Apocalisse dell'olio*. Parigi, Fayard.

Crouch, C. (2000). *Coping with post democracy*. Fabian Pamphlets

de Waal, F. (2006). *Le Singe en nous*. Fayard.

Gadrey, J., & Jany-Catrice, F. (2005). *Les nouveaux indicateurs de richesse*. La Decouverte, Reperes edition.

Gardner, H. (2006).*Multiple intelligences: New horizons*. Basic books.

Gardner, H. (2011). *Frames of mind: The theory of multiple intelligences*. Basic Books.

Gare, A. (2016). *The philosophical foundations of ecological civilization: A manifesto for the future*. Routledge.

Gist, R., & Lubin, B. (1989). *Psychosocial aspects of disasters*. Wiley.

Holmgren, D. (2010). *Permacultura. Dallo sfruttamento all'integrazione: Progettare modelli di vita etici, stabili e sostenibili*. Arianna.

Hopkins, R. (2008). *The transition handbook: From oil dependency to local resilience*. Chelsea Green Publishing.

Jackson, T. (2009). *Prosperity without Growth. Economics for a Finite Planet*. Earthscan.

Jonas, H. (1985). *The imperative of responsibility: In search of an ethics for the technological age*. University of Chicago Press.

Kallis, G. (2011). In defence of degrowth. *Ecological Economics, 70*, 873–880.

Kendra, J., & Wachtendorf, T. (2003). *Disasters: Elements of resilience after the World Trade Center disaster: Reconstituting New York City's Emergency Operations Centre*.

Kempf, H. (2010). *Per salvare il pianeta dobbiamo farla finita con il capitalismo*. Garzanti, Varese.

Kimhi, S., & Shamai, M. (2004). Community resilience and the impact of stress: Adult response to Israel's withdrawal from Lebanon. *IEEE Journal of Community Psychology, 32*(4), 439–451.

Korten, D. (2000). *The post corporate world: Life after capitalism*. Berrett-Koehle.

Latouche, S. (2004). Degrowth economics. *Le monde diplomatique*. https://www.jussemper.org/Resources/Economic%20Data/Resources/Degrowth%20economics,%20by%20Serge%20Latouche.pdf

Latouche, S. (2006). *The globe downshifted*. Le monde diplomatique.

Latouche, S. (2007). *La scommessa della decrescita*. Feltrinelli.

Latouche, S. (2011). *Come si esce dalla società dei consumi. Corsi e percorsi della decrescita*. Bollati Boringhieri.

Lavanco, G., & Novara, C. (2012). *Elementi di psicologia di comunità. Progettare, attuare e partecipare il cambiamento sociale*. McGraw-Hill Education.

Lewin, K., & Cartwright, D. (1951). *Field theory in social science*. Harper, New York.

McMillan, W. D., & Chavis, M. D. (1986). Sense of community: A definition and theory. *Journal of Community Psychology, 14*, 6–22.

Mollison, B., & Holmgren, D. (1978). *Permaculture one: A perennial agriculture for human settlements*. Transworld, Melbourne.

Morin, E. (1999). *Seven complex lessons in education for the future*. UNESCO Publishing.

Mulligan, M. (2002). *Ecological pioneers: A social history of Australian ecological thought and action*. Cambridge University Press.

Nisbet, R., & Perrin, R. G. (1977). *The social bond*. Knopf, New York.

Norris, F. H., & Stevens, S. P., Pfefferbaum, B., & Wyche, K. F., & Pfefferbaum, R. L. (2008). Community resilience as a metaphor, theory, set of capacities, and strategy for disaster readiness. *IEE American Journal of Community Psychology, 41,* 127–150.

Piketty, T. (2013). *Capital in the twenty-first century.* Harvard University Press.

Prati, G., & Pietrantoni, L. (2009). *Psicologia di comunita. Psychofenia, XII*(20).

Prochaska, J. O., & DiClemente, C. C. (1982). Transtheoretical therapy: Toward a more integrative model of change. *Psychotherapy: Theory, Research & Practice, 19*(3), 276–288.

Rappaport, J. (1977). *Community psychology: Values, research and action.* Rinehart & Winston.

Ronan, K. R., & Johnston, D. M. (2005). *Promoting community resilience in disasters: The role for schools, youth, and families.* Springer.

Shiva, V 2000. *Stolen harvest: The hijacking of the global food supply.* South End Press.

Sonn, C. C., & Fisher, A. T. (1998). Sense of community: Community resilient responses to oppression and change. *Journal of Community Psychology, 26*(5), 457–472.

Tobin, G. A., & Whiteford L. M. (2002). Economic ramifications of disaster: Experiences of displaced persons on the slopes of Mount Tungurahua, Ecuador. *Papers of the Applied Geography Conferences, 25,* 316–324.

van den Bergh, J. C. J. M. (2011). Environment versus growth—A criticism of "degrowth" and a plea for "a-growth". *Ecological Economics, 70*(5), 881–890, Science Direct, Elsevier.

World Commission on environment and development. (1987). *Our common future.* Oxford University Press, Oxford.

Zimmerman, M. A. (2004). *Empowerment theory & adolescent resilience.* European Association for Research on Adolescence.

CHAPTER 5

Ethics, Degrowth, and Resilience: The Complexity of the Transition Path Towards *DemocraCity*

During the history of mankind, several definitions of well-being have been in use. In the twentieth century, as Latouche sets out (2009) welfare was essentially considered synonymous with economic well-being, so that, after the Great Depression and the Second World War, the national economic accounts (and therefore the Gross Domestic Product) was considered by many to be the main instrument for the measurement of development.

Kuznetz, in 1934, exhorted to the Congress of the United States that the welfare of the country could hardly be inferred on the basis of the measurement of its national income only. The limits of using only one indicator in order to measure the well-being were, according to Kuznets, several: the GDP index does not include activities carried out outside the market, for instance volunteering or domestic work; the negative social and environmental externalities of the production system include the expenses for the defence, i.e. military expenses, while it does not take into account the distributive elements.

It was as late as the 1990s that initiatives dedicated to sustainable development and the measurement of human development have begun to capture media attention and play an important role on the political agenda. These included the Bruntland Report, the Human Development Index of the United Nations Development Program (UNDP), the Millennium Development Goals, or more recently, the studies on the quality

M. Carcea, *The Road Towards DemocraCity*,
https://doi.org/10.1007/978-981-16-3505-2_5

of life conducted by the Organization for Economic Cooperation and Development (OECD) by measuring social progress.

According to Giovannini (2012), due to the actions taken by local realities such as the Transition Town Movement and other national authorities, to research in relation to the quality of life, to new paradigms such as Degrowth theory and, also, to the above-mentioned research into the measurement of social progress by the Organization for Economic Cooperation and Development (OECD), a new movement is emerging aimed at measuring well-being or, trying to change the collective imaginary to develop a better understanding of the difference between well-being and well-having. Is it possible to identify, within the Transition Town Movement, such as in the example of this study i.e. Transition Darebin, an indicator that could be an expression of *buen vivir*? Will it be possible to liberate the concept of "happy" (Pallante, 2007) economic degrowth from its academic shackles and create a new useful tool for the emergence of *DemocraCity*?

The Construction of a New Imaginary

The questions raised by Latouche (2007), Pallante (2007), and Bertorello and Corradi (2011) include whether it is the destiny of the planet to become a great and unique market society. How is possible to reinvent and define the ways of embeddedness, Polanyi's definition according to Beckert (2007), of the economy in society so that human societies can exercise forms of democratic control over economic and financial dynamics? The answers to such questions are not only scientific, according to Machado (2011), as the reconsideration of the disciplinary status of economy as a discipline shows, because it must be returned to the moral sciences where it belongs and belonged, from Aristotle to classical political economy, however, the discipline characteristics of a political nature.

Let us take a moment and think about crucial issues such as the quality of democracy, the relationship between democracy and capitalism, the limits of the social-democratic compromize, the crisis of the welfare state, the question of recognizing the subaltern cultures that have been marginalized to date on the international scene, the transformation in a multicultural sense of postcolonial societies, the formation of a global civil society, and above all the ever more pressing need for supranational polyarchic governance due to the emergence of transnational challenges such as hunger, droughts, natural disasters, terrorism, and so on.

Given the scale of the unprecedented emergencies as outlined above, the narrowness of the dominant economic model seems increasingly evident. Rational action theory (RAT), which is expected to gain a reasonable following with the Chicago School (Goldthorpe, 1998) in the work of Gary Becker and Friedrich A. von Hayek, not only explains what happens in the market with the exchange of goods and money, but all forms of human action: love, crime, religious beliefs, political behaviour, education.

Such an expansion of economic science, with a variety of different approaches, anchored in intellectual contexts sometimes mutually exclusive, preceded the offensive of neoliberalism that marked the beginning of the era of so-called globalization, at least until the serious global economic and financial crisis of 2008. It is clear that the hegemony of the ideology of homo oeconomicus in the social sciences and moral philosophy related to an underlying anthropological and philosophical conception, which is what we can summarize in the term utilitarianism.

Caillé (2009) proposes an original interpretation of this doctrine that goes well beyond the canonical tale that makes Jeremy Bentham of the Principles of Morals and Legislation (1789) or John Stuart Mill of Utilitarianism (1861) the founding fathers, and moralists Scots D. Hume, A. Smith of Helvetius and Beccaria the forerunners. Caillé, who carries the legacy of Mauss and Polanyi, pioneers of the économie solidaire, distinguishes between a broad and narrow meaning of the doctrine of utilitarianism. Utilitarianism designates all those theories that identify the greatest possible happiness with the individual interest (individual self-interest), obtained through contract and the free market.

With the marginalist economy, starting in the 1870s, the ascendance of the hegemony of utilitarianism began. It was understood as extended practice rationality, that is, its postulate of the rational egoist subject progressively tended to conquer all areas of the human and social sciences, moral and political philosophy in their entirety, reaching an almost uncontrolled predominance of RAT. Utilitarianism thus becomes not only the dominant theoretical and ideological system, but the very imagination of modernity, the faithful mirror of the anthropology of modern man in his reduction to homo oeconomicus: a triumph that celebrates its glory in today's era of so-called globalization.

What is this unprecedented phenomenon that is globalization? According to Caillé (2009), it is the subordination of all spheres of social existence to the law of the market alone, to be precise primarily the speculative capital market. Globalization of this king subjects all spheres of social action to itself, (science, culture, technology, sport, and so on) everything becomes a commodity. The principle of utility, according to Arendt (2004), reaching its extreme consequences, breaks the relationship between man and worldly things, which are no longer considered in their objective utility, but simply as more or less casual results of the production process in view of subjective "satisfaction" through consumption. While *homo faber*, according to Arendt, was interested in the final product with its particular use value, and which the whole production process was geared toward, the utility principle is primarily concerned with stimulating productivity and reducing suffering.

The true unit of measure is not utility nor use, according to Arendt but satisfaction, that is, the quantity of pain and pleasure experienced in the production or consumption of things. Generalized utilitarianism leads to the victory of the *homo laborans*, which is the individual projected on himself, projected into the closed interiority of introspection, whose only remaining contents are the appetites and desires, the unconscious needs of his body that he declared passions and that he judged unreasonable because he realized that he could not reason, that is, to reckon with. It is the victory of the consumer society, whose dominant ideology is aimed at promoting the vital process of the species, mankind as an animal species.

The Gift: An Introduction

According to Caillé (2005), the breakthrough achieved by the marginalism economic revolution announces, therefore, that any good or service that is desired is useful provided there is a means to pay for it: to decide on the usefulness of things is the ability to buy them (money) only, which then becomes the exclusive criterion for measuring wealth and, therefore, the happiness of individuals and nations. The consumer society, according to Caillé (2005), is a *phantasmatic* subject, de-realized, devoid of the common world, and prey to the logic of unlimitedness and excess (hubris) in its various forms and modalities (financial, ecological, biotechnological, criminal, etc.), which knows no obligations to others and no reciprocity.

Using this perspective, Robertson argues (2012), the researchers gathered around the MAUSS, (acronym for Mouvement anti-utilitariste dans les sciences sociales)—even among the inevitable differences that exist between them—will qualify as utilitarian every purely instrumental conception of existence, which organizes life in the function of a calculation or a systematic logic of means and ends, for which action is always performed in view of something other than itself, and ultimately referred back to the individual. A subject who is assumed to be closed off from other people, a master of his own actions, recipient and beneficiary of doctrines for which the *interests for*, passions, emotions are or should be *interests to*: useful passions.

The representatives of the MAUSS consider it an urgent task to elaborate an anti-utilitarian conception of existence; and to this end propose for everyone, using different discursive methods, to reconnect with the French anthropological tradition of Émile Durkheim and Marcel Mauss. A particular focus is to develop the paradigm of the gift, whose features were sketched by Mauss in his famous *Essai sur le don*. Caillé, before starting his reinterpretation of the gift theory, starts from the original philosophical questions about moral problems: what is a fairer society? How should we live? What are our social obligations? What is democracy? What is happiness? How to define wealth?

Framing the gift paradigm as the focal point of an anti-utilitarian conception of existence does not at all mean denying the legitimacy of one's own interests, personal wealth and happiness, the existence of other passions, which push the individual in the direction of the interest of others, of compassion and the well-being of society. Caillé (2005), starting from the ultraliberism that regulates the dynamics of *homo oeconomicus*, points out the distinction between interest for (the proving of interest for others or the passionate interest in certain activities and for certain knowledge: for mathematics, literature, sport, etc.) and interest in (the instrumental interest in achieving a professional career, the strategic interest, selfish interest or for oneself or the interest to passively obey the will of others). He also adds the distinction between disinterestedness (i.e. total absence of interest in taking an action) and disinterest (an action taken beyond an instrumental and selfish interest), making Adam Smith's lesson into his own.

Starting with Smith's belief in the universal human ability to feel compassion, Caillé argues that moral sense and a sense for community are the empirical bases, the tool to realize an anti-utilitarian theory. Caillé coins a neologism, *aimance*, to indicate openness and solicitude towards the other, which is a starting point of action as primary and irreducible as "self-interest", moral "obligation" and "freedom". Within the concept of aimance, the author includes friendship, philia, agape, caritas, piety, solidarity, altruism, cooperation, alliance, association, all those behaviours inspired by reciprocity. Gift and interest are intertwined in the practice of men: *interest for* and *interest to* intersect and overlap.

Starting from the basis of Mauss's teaching, not only in archaic but also in modern societies the spirit of the gift is not at all identifiable with the unilateral gift and without reciprocation. A unilateral gift, combined with an attitude of superiority on the part of the donor, is humiliating for the recipient. According to Mauss human beings are a two-faced Janus: the conflict with his fellow men can be resolved in a peaceful agreement or result in massacre. Just as the gift can be positive or negative, bearer of death (of offense, humiliation, contempt) or of salvation (alliance, pacification), an ambivalence that Mauss found in the word "Gift" of the Germanic languages, present at its turn in the Greek word *pharmakov,* which means at the same time the gift of the doctor, the medicine that heals and the poison that kills.

This semantic duplicity is found in the Greek word *dosis*: something that kills or can heal according to the dose. Caillé argues that these two opposite meanings refer to the polarity between self-interest and interest for the other, to the primitive inability to dissociate these two interests, and to the reversibility of self-interest and interest for the other. From this perspective, the agonistic dimension of the gift, emphasized by Mauss, is extremely important, since it allows human beings to form an alliance and establish a genuine social relationship. The desire to appear (*Selbstdarstellung*) is the essence of the *homo politicus* and, what matters most, it is a tendency to be found throughout the animal world (and even in the vegetable world). Arendt (1987) emphasized, that all living beings, men and animals, not only are in the world, but are of the world, and this is precisely because they are both subjects and objects, in which they perceive and are perceived.

Appearing for human beings, according to Robertson (2012), is equivalent to entering the order of narrativity, since human action is by definition unpredictable and creator of meaning: through human action, which breaks the deterministic chain of causes and effects by introducing to the world always something new, the identity of the subject becomes, from a physical identity that was, a narrative identity, endowed with a singular and unrepeatable history. Caillé conjugates, according to Robertson (2012) in the same discursive register, the Arendtian concept of action and the Maussian notion of gift.

Both converge in emphasizing the dimension of meaning, of the possible and of the new respect for the established order, and e socially and culturally consecrated meanings. The value of the subject is proportional to what he "gives" or, what is the same thing, to the underlying reasons/causes of his actions, if we recall the fact that there is a multiplicity of manifestations of action: from the gratuitousness of donation, found for instance, in beauty or inspiration, to the effort and punishment which, in terms of work, is the cost the production of an object: from love, charity and compassion towards others to the innovation and creativity of a work of art.

If, as Robertson (2012) has argued, the fundamental issue for human beings is to appear, then the goal of recognizing the value of existence (for communities and/or individuals) becomes a cornerstone. The desire to appear, continues Robertson, in human beings, is one with the desire for recognition, to the point that it is legitimate to affirm that men aspire more to be recognized beings than to accumulate wealth. Within the *potlàc* of some archaic societies, Mauss identifies the keystone that holds together the recognition of authority, of prestige, of social esteem on the one hand, and the circulation and redistribution (key concept within the broad framework of degrowth and the Transition movement) of goods on the other. Recognition, rooted in the order of culture, and redistribution, rooted in the economic order, are combined due to the contribution of Caillé.

The gift is the keystone in the production of the social bond and establishes the space of the political. Caillé emphasizes that the *potlàc* is a fight for generosity. This means that recognition in archaic societies is related to the obligation of the gift framed as a redistribution. The system of recognition/redistribution, according to Robertson (2012) always brings with it an element of gratitude, reflecting the fact that the other side of giving oneself to others is a collective, or positive indebtedness. What we are we owe to the inspiration, to the generosity of others. The obligation to receive is a social imperative: in archaic societies rejecting a gift means expressing the fear of not being able to reciprocate, of not feeling equal; the fear is to lose the esteem of the community. However, according to Mauss (2002) accepting a gift is equivalent to showing the certainty of being able to reciprocate and, therefore, of being able to prove that one is not inferior.

The circulation/redistribution of things is related to the circulation/recognition of rights, privileges, and authority. What circulates and is exchanged are objects of recognition and, as Robertson has argued (2012), of relations of recognition, of which they are integral parts of power relations, relations between rulers and the governed. As Philippe Chanial (2008) argued, studying society with the filter/lenses of the gift paradigm implies as a consequence "jointly thinking, in the same framework of analysis, material circulation and circulation that we can define symbolic since looked in optics of the gift, every relationship is mediated by symbols and it is these symbols (*sumbolon*—signs of recognition) be it a word, a gift, a handshake, that seal the alliance" (p. 33).

It is this balance between material dimension and symbolic dimension that allows us, according to Robertson (2012), to discover the dynamics of the gift even within the market, since even the market, as Chanial (2008) emphasizes, is a regime of social relations that, like all other modes of interaction, does not end in the axiomatic field of the interests, but presupposes as primary the rule of the obligation to donate with respect to the norm of reciprocity of reciprocating.

The Gift of Active Citizenship: Between Politics and Responsibility

The complex structure of the gift paradigm, such as it being a "game" of redistribution/recognition, also makes it a main character in the expression of politics. Because the stakes of the cycle of donate/receive/reciprocate is the recognition/redistribution of the statutory functions of a social system or of a community (and of the subjects connected to these functions), the Maussian Gift theory allows us to leave behind the contractualist fictions of modern political philosophy and the aporias of the functionalistic and systemic theories of contemporary social sciences.

Politics, according to Mauss (2002) is a form of gift, if by this expression we mean the passage from hostility to alliance, from war to peace, from predation to trade and, therefore, to the intertwined game between recognition and redistribution. This entails the decision to come to terms, to live collectively, to establish institutions in which, as Mauss explains, we can oppose without obliterating, and give oneself without sacrificing oneself to one another. Politics is, as Mauss has argued, a founding gesture that is renewed in the daily practice of social relations, where, instead of gifts, we can also exchange evils, revenge, humiliation, thus replacing the positive cycle of giving/receiving/returning with the negative cycle of taking/rejecting/keeping for oneself.

According to Robertson (2012), this is the ambivalence of the gift that is one with the ambivalence of human relations. To eliminate it, it is not enough to invent devices, orders, universal, and impersonal procedures, as was the case in modern society with the establishment of the market, the construction of the state, the imposition of the sovereignty of the law and so on: all those aspects of modernity that Weber summarized in the concept of Western rationalism. These institutions, which Caillé (2009) situates in the domain of secondary sociality, where the functions performed by people are less important than their personality, because they obey the imperative of efficiency, cannot erase the relations of primary sociality, those relationships that are formed in the family, the neighbourhood, friendship, love, in small associations, or communities.

According to the French author in the primary sociality, we remain bound to the cycle of giving/receiving/reciprocating, and on which the "vital worlds" that grow and reproduce in it, functional systems governed by the imperatives of efficiency and the corresponding communication codes are founded. As Caillé underlined (2009), functional systems could not be effective if they did not support the networks of relationship between person and person, different forms of primary sociality, inherited or built, mobilizing loyalty and the desire to give oneself to the members of this primary sociality. Caillé argued that the networks of primary sociality can exist only if incorporated into more general orders that ensure not only material, but symbolic and narrative reproduction.

Caillé, inspired by the work of Claude Lefort, who criticized the increasing depoliticization that the social sciences during the twentieth century and who was convinced that a virtuous circle should be established between philosophy (in particular, political philosophy) and the social sciences that confers the ability to speak about and to today's polis with each other. Caillé believes that politics is the outcome of the decisions through which everyone gives (or refuses) to everyone, without giving himself to anyone in particular. Politics thus appears to be the historically variable result of the ambivalence of the gift, of the triple obligation of giving/receiving/reciprocating, which is the universal matrix of all human sociality, and that in the intertwining of positive gift and negative gift, between the production of sense and redistribution of goods, it is common one shared by the most advanced and less evolved societies.

In such ambivalence, Caillé argued (2009) that its roots lie the decision to live together, once again taken by everyone and by anyone, which promotes a common history, not without conflicts and discord, through which the between family and strangers is formed, between friends and enemies, (and between they and us) and, whose recognition passes through a story to several voices of the subjects (not only individuals, but also ethnic groups, social classes, etc.) that are an integral part. Especially in contemporary democracies, whose promise consists in providing the greatest number of the greatest power to live and act or, according to Amartya (1982), equal chances to develop the capabilities of each, the recognition in question constitutes a sort of foundational act that must be retrospectively renewed every time, generation after generation.

Caillé and Chanial (2008) argued that the roots for the decision to live together can be found in the ambivalence of the gift because, again, it is a decision made by everyone, creating a common history. This history is not created without conflict or discord which creates the division between family and strangers, friends, and enemies (and between them and us). This history is created and recognized through stories told by many voices (not only individuals but also ethnic groups, social classes, and other groups) and those stories and voices are an integral part of the history. Especially in contemporary democracies, whose promise consists of providing the greatest number of people with the greatest power of agency or, according to Amartya (1980), equal opportunities for every individual to develop their capabilities, the recognition of such belonging constitutes a sort of foundational act that must be retrospectively renewed every time, generation after generation.

Even more so today, as Fistetti has argued (2008) in the age of globalization, in which societies become structurally multiethnic and multicultural and cultures, once marginalized or devalued, require to be appreciated in the irreducible singularity of their way of inhabiting the world. This reciprocal recognition produces, according to Berthoud (2005), the transition from legality to legitimacy as the most important act of political innovation, which cannot but drag with it the re-evaluation and/or redefinition of the criteria and contents of the redistribution in the polygamy of its forms (market, state, solidarity economy, etc.). A passage, stressed by Berthoud (2005) from legality to legitimacy that is made even more necessary by the emergence of a virtual world society that overlaps with the networks of primary sociality and the impersonal institutions of secondary sociality, but in which risks are being modelled exclusively by the logic of generalized commodification.

Here is an unexplored field of research, according to Fistetti (2008) around the question of how the obligation to give/receive/reciprocate is articulated within a planetary society that is forming in the interstices of national societies not only through globalist dynamics of the market, but also through the construction and dissemination of knowledge and multicultural and intercultural practices, where intertwined questions of recognition by peoples and cultures and questions of global justice arise. The *vexata quaestio* that stands out against such background is whether and to what extent it is possible, paraphrasing Polanyi, to reframe the uncontrolled beast of the capitalist economy within democratic governance.

It is a twofold question: the first is strictly economic, because it calls into question the construction of a plural economy, such as to go beyond the classical dichotomy between market and state and put into motion a mixture between the latter and the gift position (that is to say, the social economy and reciprocity such as the Transition Town Movement, the non-profit sector and so on). The second is political in the true sense of the word, because it requires the reconsideration of the modern project and the philosophical-political categories embodied in it, such as the state, nation, sovereignty, popular will, equality, individual rights, and so on; the foundations for an inclusive and sustainable democracy.

In the pursuit of the road to *DemocraCity*, the Maussian conception of the gift can provide valuable theoretical and normative indications, if it is true that citizenship becomes the primary asset of globalized societies, around which, together with the key theme of work, social conflicts are polarized, and for the definition of a fairer society. The gift of citizenship to the other: this is the stakes of contemporary democracies, knowing that the gift is a mix of unconditionality and conditionality, of selflessness and self-interest, of freedom and obligation. In the sense, to be clear, that even in the ethic of welcome and hospitality, the material advantages of the host cannot be included, as well as hostility and conflicts to be solved with compromize, agreement, negotiation. The friend/foe's relationship, paraphrasing C. Schmitt must be overcame with the concretization of Kelsen's *Civitas Maxima*.

There is no doubt according to Caillé (1998), that there is a twofold correspondence between social citizenship and political citizenship, between political citizenship and labour status (recognition of the ethical-social value of benefits, salary, social security protections, and so on). Politics in this sense, in the sense of the noblest of all the arts, is the basis of living together, it is a harbinger of that interweaving of self-interest and generosity, between conflict and association/alliance, between redistribution and recognition.

Ethics Within *DemocraCity*: The Lesson of Edgar Morin

Nature and ethics. The juxtaposition of nature and ethics while intriguing raises some questions. Ethics, in the long history of research in the field that sees the Transition Darebin community as a protagonist, plays a fundamental role in the creation of a society based on degrowth. Complex philosophical thought plays a seminal role helping us understand the meaning and application of ethics. Descartes, considered a pioneer of modern philosophy of nature, refined and reinterpreted the concept of Naturalism, established by the Mileto School, in VII sec. BC, and during the seventeenth and eighteenth century. an author. On the basis of his metaphysics, on the one hand, and physics, on the other, Descartes elaborates a conception of man diametrically opposed to that of the Aristotelian tradition and returns to a certain extent to the Platonic conception.

In fact, according to Aristotle, man is a single substance whose body constitutes matter and the soul form, while, in line with Platonic thought, Descartes man sees man made of two substances, each existing independently of the other. In Cartesian anthropology, the soul does not only indicate a spiritual principle, but a real "thinking substance (*res cogitans*), created directly by God independently of the body, equipped with a free and immortal will, and therefore independent of the determinism of natural laws that is. The body, which is instead "merely extended substance" (*res extensa*), comparable in all respects to a machine (or *autòma*), functioning according to the laws of nature and therefore mortal. Such distinction constitutes the Cartesian dualism synthesized in the masterpiece "*cogito ergo sum*".

Cartesian dualism represents modern anthropocentrism strengthened by idealism of Hegel, who considers nature as evil, alienation and incidental. Accordingly, Marx conceives nature as an instrument to enhance man's work and creativity; both framed nature as an element that must be dominated by man in order to fulfil himself. Within the philosophy of western cosmology, there are also counterparts, theories that empower nature as spirit and evolution, however they are not part of our imaginary yet. I refer, for instance, to the school of trans-disciplinarists, in which Morin is viewed as one of the founders of this field of research.

With his *Seven complex lessons in education for the future,* Morin (1999) offered an invaluable contribution regarding the imperative of responsibility, ethics and nature reversing the above conception, basis of western philosophy. Morin enucleates and explicates (1999) seven topics that need to become essential in teaching and education and should be included in every society and in every culture. Paraphrasing Morin, these themes will allow us to integrate the existing disciplines and to stimulate the development of a knowledge able to meet the challenges of our individual, cultural and social life.

This is not the place to analyse all the contents of the work written for UNESCO, however it is absolutely necessary for me to use the work, in this case the fourth, fifth, and seventh chapter of *Seven complex lessons in education for the future*, to create a "an ethical constitution" of what will be my structure of *DemocraCity*, the soul of this long journey of study. Sharing the ideas of Latouche, Illich, Boulding, and other scholars, fundamental to this research journey, Morin concludes the fourth chapter hoping for an awareness of the human condition and the birth of a new awareness that sees man as citizen of the earth, with so, he intended to teach an earth identity.

Globalization, according to Morin, is the emergence of a new subject: the world as such. In the age of telecommunications, the individual is overwhelmed by the complexity of the world and information suffocates our capabilities for making the world intelligible. The difficulty of knowing our world is exacerbated by our current way of thinking that has atrophied our ability to contextualize and globalize, while the need of the planetary era is to understand globality, multidimensionality, complexity. It is the productive and destructive complexity of the reciprocal actions of the parts on the whole and on the parts that constitutes a problem. The planet is not a global system, but a moving vortex. It requires, the philosopher argues, a polycentric thought.

The history of mankind began with a diaspora on all continents; then, in modern times, it entered the planetary age of communication among all the fragments of the human diaspora. It has not produced genetic cleavages, but an extraordinary diversity of languages, cultures, and destinies. To paraphrase Morin (1999), the treasure of humanity lies, in its creative diversity, but the source of its creativity is in its generation of unity. The planetary era opens and develops through violence, destruction, slavery, and fierce exploitation. Planetarization develops on all continents with the contribution of European civilization, its weapons, technology, and conceptions. Economic development and communications lead to formidable migratory flows. Planetarization has given rise to two wars and two global economic crises and, after 1989, to the global spread of the liberal economy.

From One Crisis to Another: Recurring Cycles in the History of Civilization

The world economy is increasingly an interdependent whole. While Europe lives in its comfort circuit (an assumption Morin shares with Latouche, Shiva, Kallis, and others), a great number of Africans, Asians, South Americans, are in a misery circle and suffer the repercussions of the world market influencing the prices of goods produced in their countries. Each part of the world, according to Morin, is increasingly part of the world and the world as such is increasingly present in each of its parts. This occurs for nations and peoples, but also for individuals. Globalization is unifying, but it is also conflictual in its essence. The globalizing unification is more and more accompanied by the negative that it produces as a counter effect: balkanization.

The antagonisms between nations, religions, between secularism and religion, between modernity and tradition, between democracy and dictatorship, between rich and poor, between East and West, between North and South, feed on each other, meeting in areas of interference and fracture. Thus, according to Morin, the twentieth century created and divided a single planetary fabric; its fragments are isolated, stiffened, and fight among themselves. At the same time, the technical-industrial irruption on the globe tends to suppress many human, ethnic, and cultural differences. Development has created more problems than it has solved. We need a richer and more complex concept of development, which is material, intellectual, emotional, moral at the same time. The twentieth century has not come out of the planetary Iron Age, but as Morin has argued, it has sunk into it.

The twentieth century has made enormous progress possible, but it was also the century of the alliance between two barbarities. The first comes from the dawn of time and brings war, massacre, deportation, and fanaticism. The second one, framed by Morin as a cold, anonymous force, is intrinsic to a rationalization that knows only calculations and ignores individuals, their flesh, their feelings, their souls, multiplying the powers of death, and techno-industrial subjugation. To overcome this barbarous era, it is necessary, according to Morin, to recognize its double legacy: of death and birth. Morin takes the lesson of Kafka: the evolution of mankind is a growth of the power of death. Since the twentieth century, it has also been that of two new powers: the global death of all humanity through nuclear weapons. The march of humanity is now accompanied by the potential for self-annihilation?

According to Morin (1999), the second is the possibility of ecological death. The unrestrained domination of nature through technology leads humanity to suicide. Finally, death has gained ground within our souls. The powers of self-destruction, latent in each of us, have been activated (also with drugs), where solitude and anguish multiplies and grows. Civilization born in the West believed in heading towards a future of progress indefinitely. We have understood, as Morin has argued, with Hiroshima, that science is ambivalent. If modernity is defined through unconditional faith in progress, in technology and in science, in economic development, then this modernity is dead. However, according to Morin, it is still possible to glimpse a hope, the possibility of terrestrial citizenship, a long and arduous path towards what we call *DemocraCity*.

The twentieth century has inherited some regenerating counter currents. In particular counter currents such as ecological, qualitative, resistance to a utilitarian life (devoted to love, enchantment, passion, festivity), resistance to the primacy of standardized consumption (manifested in the search for a lived intensity or frugality), f emancipation from the tyranny of money, the pacification of souls and minds. One of the fundamental conditions for a positive evolution is that the emancipatory forces inherent in science and technology can overcome the forces of death and enslavement. Technological developments are ambivalent. Humans subjugate machines, which subjugate energy, while they are themselves subjected to it. The possibilities offered by biology are equally prodigious in the positive and in the negative way.

The multitude of the transformations seems to announce, as Morin has argued, a mutation even more considerable than that of the Neolithic Age, caused by the transition from hunter-gatherer societies without state, agriculture, and city, to the historical societies which have spread across the planet for eight millennia. We can also count on the inexhaustible sources of human love and hope for the intellectual possibilities of humans, which are still partly unexploited. Since social possibilities are related to the intellectual ones, nothing can guarantee that our societies have exhausted their possibilities for improvement and transformation. The anthropological, sociological, cultural, spiritual possibilities of progress restore the principle of hope, but without scientific certainty, nor historical promises. It is an uncertain possibility, which depends a lot on awareness, on will, on courage, on luck; only in this way the awareness, Morin concluded, becomes urgent and primordial.

The planetary union, the minimal rational requirement of an interdependent world, needs a conscience that binds us to our Earth considered as the first and last homeland. According to Morin, we must learn to "be" on the planet: learn to live, share, communicate, and be in communion. We no longer have to be just a culture, but also terrestrial. We must inscribe in us the anthropological consciousness (unity in diversity), ecological (inhabiting, with all other mortals, the same sphere of living; and abandon the Promethean dream of the domination of the universe to nurture an aspiration to conviviality on Earth), civic (responsibility and solidarity), and dialogic (criticism, self-criticism and understanding). We must cease to construct an opposition between the universal and the homelands and start to link our homelands concentrically to integrate them into the concrete universe of the terrestrial homeland. The search for a better future must be complementary, not antagonistic, with a return to the origins of the past.

Every human being, continues Morin, collectively must enrich his life with a constant circulation between the past, in which his identity is rooted, the present, in which he affirms his needs, and a future, into which he projects his aspirations and his efforts. States can play a decisive role, provided they abandon absolute sovereignty. The era of fertility of the nation states endowed with absolute power has ended. The Confederate world must be polycentric, even culturally. Reliance must replace disjunction and rely on the wisdom of living together. Mestizo becomes product and producer of reliance and unity.

The mestizo can develop a poly-identity starting from his family's multipolarity (ethnic, national and on) and constitute in himself a fully human complex identity. We are committed to continuing the essential work of life which is to resist death. The consciousness of our humanity should lead to solidarity and mutual sympathy. Morin argued that education must include an ethic of planetary understanding. Morin urges future generations to expect the unexpected: previous centuries believed in a repetitive or progressive future. The twentieth century has discovered the loss of the future, its unpredictability. The greatest achievement will be to eradicate the illusion of predicting human destiny.

The future remains open and unpredictable. The awareness of historical uncertainty comes to pass in the collapse of the myth of progress. Progress is possible, however, according to Morin, uncertain as the unpredictability of the events of 1914, 1917, 1929, 1939, 1989, and of course, the global financial crisis of 2007 shows; the becoming is now problematized and will be forever. The future is called uncertainty. History, according to Morin, advances starting from local and almost microscopic creations. Despotism and totalitarianism know that deviant individuals constitute a potential threat: they eliminate and annihilate the micro-outbreaks of deviance.

Every evolution, the philosopher continues, is the result of a successful deviance, whose development transforms the system in which it was born: deviance disorganizes the system while reorganizing it. In addition to innovations and creations, there are also destructions. There is always a great dispersion of human experience acquired from previous generation. History, according to Morin is not a linear evolution: it is a crossover of conflicting processes, with risks and uncertainties, which entail evolutions, involutions, progress, regress, and breakdowns. It always has two sides: civilization and barbarism, creation and destruction, genesis and death.

Long Term Unpredictability: From Pascal's Lesson to a Desirable Strategy

In the fifth chapter of his work, Morin insists that to a universe obedient to an impeccable origin, we must replace a universe that is the game between order, disorder, and organization. We must learn to face uncertainty. Education must recognize the uncertainties related to knowledge because they are worth uncertainty: cerebral-mental (derives from the translation-reconstruction process of all knowledge), logic (neither true nor false), rational (rationality can result in rationalization), and psychological (we are not fully aware of what is happening in our mind). According to Morin, dogmatic and intolerant certainties breed the worst illusions.

Awareness of the uncertain character of cognitive acts is an opportunity to arrive at a pertinent knowledge, which requires examination, verification, and convergence of clues. Life involves cases without definition or with false definitions and, above all, the absence of a closed general framework. Our reality, continues Morin, is only our idea of reality. Therefore, it is important not to be realistic or unrealistic in a banal sense, but to be realistic in the complex sense of the term: to understand the uncertainty of reality, to know that reality includes a still invisible possibility.

From the moment in which an action is undertaken, it begins to escape the intentions. It is necessary to take into account the complexity that the action entails, with its risks, initiatives, decisions, and unexpected events. The ecology of action involves four principles of uncertainty:

- The risk ↔ (pre)caution loop: for every action taken in an uncertain environment, there is a contradiction between the risk principle and the precautionary principle.
- The ends ↔ means loop: the means and ends retro act on each other and it is inevitable that ignoble means at the service of noble ends pervert the latter and end up replacing the ends.
- The action ↔ context loop: every action escapes the will of its author by entering the game of inter-retro-actions of the environment. This can have three types of unexpected consequences: the perverse effect (unexpected nefarious effect, more important than the expected benefit), the innovation of innovation (the more it changes, the more it remains the same), the putting at risk of the innovations achieved.

According to Morin, there are two main cornerstones: the first is the full awareness of the wager that the decision entails, the second is the recourse to strategy. Pascal's wager must be generalized to every faith. Strategy must prevail over programme. It elaborates an action scenario examining the certainties and uncertainties of the situation. In an unstable environment, strategy must sometimes favour caution, sometimes audacity, or both at the same time. A strategy at the service of maximum "freedom, equality, fraternity" (Morin, 1999, p. 47) is difficult. These complementary terms are at the same time antagonists. Strategy, like knowledge, is navigation in an ocean of uncertainties, through archipelagos of certainties.

The desire, as Morin suggested, to eliminate uncertainty can then appear to us as a disease of our mind, and every path to great certainty could only be a hysterical pregnancy. Thought must arm itself to face uncertainty. Everything that involves possibilities involves risk and thought must recognize the possibilities of risks as well as the risks of possibilities.

Morin concludes his fifth chapter with this expression of hope:

> Giving up on progress guaranteed by the "laws of History" does not mean giving up on progress but recognizing its fragile uncertainty. Renouncing the best of all worlds does not at all mean renouncing a better world.
>
> Alas we have often seen the possible become impossible in history, and we have a premonition that the richest human possibilities are still unrealized. But we have also seen the unhoped-for become possible and fulfilled; we have often seen the improbable come true instead of the probable. Let us know how to hope for the unhoped-for and strive for the improbable. (Morin, 1999, p. 48)

Teaching and the Value of Understanding

It is undeniable how Morin refers to the lesson of the limits of growth and Gift theory at the beginning of the sixth chapter of Seven Complex Lessons. On our Earth, the French philosopher argued (1999) interdependencies have multiplied. The awareness of being in solidarity with life and death should now tie humans to each other, but misunderstanding remains general. The problem of understanding has become crucial. Understanding should not be quantified. Educating to understand mathematics or another discipline is one thing; educating human understanding is another.

The problem is doubly polarized: at one extreme of the scale we find the planetary and the other the individual. While it can be argued that the closer one is to an issue, the better one understands, this is only a relative truth and the opposite can also be argued, i.e. the closer one is, the less one understands. Proximity, according to Morin, can feed misunderstandings, jealousies, aggression, even in apparently more evolved intellectual environments. Communication, continues the author, does not produce understanding. Information produces intelligibility, a necessary condition, but not sufficient for understanding. There are two levels of understanding: intellectual or objective understanding and inter-subjective human understanding. Explain is regarded as an object that you must know. Explanation is necessary for intellectual or objective understanding. Human understanding goes far beyond: understanding involves a process of empathy between subjects, identification, and projection. "An ego alter that becomes an alter ego", here Morin (1999, p. 50), reverses Schmitt's friend/foe theory in order to embrace the concept of Civitas Maxima of Hans Kelsen (1961): thus, the international community is the personification of the world or universal legal order. Kelsen insists on the need to conceive of the latter as a real state, such as the world or universal state, the *civitas maxima*. Understanding the meaning of another's words, his ideas, his vision of the world is threatened by noise, by polysemy (a concept used in one sense is understood in another), the ignorance of rites and customs of others, by misunderstanding of the values of another culture, by the impossibility of grasping ideas or arguments of another world view, by the incommensurability between one mental structure and another.

Egocentrism. It maintains, according to Morin, self-deception, generated through self-justification, glorification, and the tendency to project the cause of all evil on others. Self-incomprehension is a very important source of misunderstanding of others. The individual's own shortcomings and weaknesses are concealed from themselves, which makes them ruthless towards the deficiencies and weaknesses of others.

Ethnocentrism and socio-centrism. They, according to Morin, nourish xenophobia and racism, which can go so far as to deny the foreigner the quality of being human. Indignation indicates a shortfall in self-analysis. The moral disqualification avoids any effort to understand … a moral refusal expresses the refusal to analyse and think.

The reductive mind. Reducing the knowledge of complexity to that of one of its elements has worse consequences in ethics than in physics. This type of knowledge leads to the reduction of a personality to one of its traits. If it is favourable, there will be a misunderstanding of the negative aspects of the personality. If unfavourable there will be mismatch of the positive traits. In both cases, there will be misunderstanding.

Ethics of understanding. In the final part of the sixth chapter, Morin offers several insights how to trace a route towards *DemocraCity*. Ethics of understanding is, according to Morin, an art of living that requires great effort to understand disinterestedly and cannot expect any reciprocity. The ethics of understanding requires understanding misunderstanding.

Thorough thinking. It allows us to learn the text and the context, the being and its environment, the local and the global, the multidimensional; basically, the complex.

Introspection. Self-examination of self. If we discover that we are all fallible, fragile, insufficient, caring, then we can discover, according to Morin, that we have a mutual need for understanding. Critical self-examination allows us to recognize and judge our self-centeredness.

Awareness of human complexity. Understanding of others requires awareness of human complexity.

Subjective (sympathetic) open-heartedness to others. Morin uses the example of the cinematographic medium (he cites Chaplin and his mask, the trump). He stresses that such medium allows us to sympathize with and understand those considered strangers or unpopular, while in real life we are almost indifferent.

Tolerance interiorized. True tolerance is not indifference. It presupposes a conviction, a faith, an ethical choice, and the acceptance that ideas contrary to ours are expressed. It involves suffering in tolerating that harmful ideas are expressed. There are four degrees of tolerance:

1. respect the right to make a speech (lesson that Morin has learned from Voltaire);
2. respect for antagonistic ideas (inseparable from the democratic option);
3. the conception that the opposite of a profound idea is another profound idea (respect the truth inherent in the antagonistic idea of ours (lesson that Morin learned from Niels Bohr);
4. the awareness that humans are possessed by myths, ideologies, ideas, or gods, and can be carried away in directions they had not intended to take.

Morin concludes his sixth chapter proposing a teaching methodology based on an inter-poly-trans-disciplinarity that helps the formation of a well-made mind capable of that complex thought adequate to the understanding of the dynamic needs of planetary interdependence. The ethics of understanding between people, according to the French author, must be connected to the ethics of the planetary age that requires globalization of understanding. Cultures must learn from each other: in the case of fine arts, music, literature, thought, cultural globalization is not homogenizing. Western culture can appear to other cultures at the same time not comprehensive and incomprehensible. However, the open and self-critical rationality born from a European heritage allows the understanding and integration of what other cultures have developed.

Understanding between societies presupposes open democratic societies; but, in addition, there remains the epistemological problem of understanding: in order to reach an understanding between thought structures, one must acquire a thoughtful meta-structure that understands the causes of one's misunderstanding with respect to others and which can overcome them.

> Understanding is both the means and end of human communication. Our planet needs mutual understanding in all direction. Given the importance of education for understanding, on all educational levels and for all ages, the development of understanding demands a planetary reform of mentalities: this is a task for education of the future. (Morin, 1999, p. 55)

Ethics for Future Generations

Individuals are more than just products of the reproductive process of the human species, because this process is produced by individuals. Interactions between individuals produce society and this feeds back on individuals. Thus, individual ↔ society ↔ species are not only inseparable, but co-produce with each other. Each of these terms, according to Morin is, at the same time, the middle and the end of the others. One cannot make only one of them the supreme end of the triad: this is itself in its own rotary end. Consequently, these issues cannot be dissociated. *Anthropoethics,* according to Morin, presupposes the conscious and enlightened decision to: assume the human condition (Morin has framed it as a triad, see Chapter 3 of his *Seven Complex Lessons*) in the complexity of our being: to realize humanity in us, in our conscience; to assume human destiny in its antinomies and in its fullness.

The anthropological mission of the millennium, according to the French author is:

- to work for the humanization of humanity.
- Obey life and guide it.
- To realize planetary unity in diversity.
- Respect the difference and self-identity in others.
- Develop the ethics of solidarity and understanding.
- Teaching ethics for the mankind.

Anthropo-ethics, concludes Morin, is individual consciousness beyond individuality.

A Laboratory of Ideas for Democracy

Individuals and society exist as mutual structures. Democracy allows their rich and complex relationship. Democracy is based on the control of the power by the controlled, reducing enslavement. Learning from Hans Kelsen (1964), Morin argues that authoritarian or totalitarian societies colonize individuals; in democracy, the individual is a citizen, a person who expresses his desires and his interests, is responsible and in solidarity with his city. Democracy cannot be defined in simple ways. The sovereignty of the people entails self-limitation through obedience to laws and the transfer of sovereignty to the elected. At the same time, it involves the self-limitation of the influence of the state, through the separation of powers, the guarantee of individual rights and the protection of private life.

The experience of totalitarianism has highlighted a fundamental character of democracy: its vital link with diversity, just as the diversity of species must be protected to safeguard the biosphere. Democracy presupposes and nourishes the diversity of interests as well as ideas. It cannot be identified as the dictatorship of the majority over minorities. It needs conflicts of ideas and opinions; they give it a vitality and productivity that can only manifest itself in obedience to the democratic rule.

Claiming at the same time consent, diversity, and the coexistence of conflicting ideas, democracy is itself a complex system of political organization and civilization: it nourishes and nurtures the intellectual autonomy of individuals, their freedom of opinion and expression, their civic sense; it nourishes and feeds on the ideal of Liberty ↔ Equality ↔ Fraternity, which involves a creative conflict between the three inseparable terms. Morin teaches us that the development of political, economic, and social complexities fuels the development of individuality and, through such development, individuality asserts itself in its own right and acquires existential freedoms.

The cornerstone of the final chapter of Morin's lesson is the ***democratic dialogic***. All the important features of democracy have a dialogic character that unites complementary antagonistic terms: consensus and conflict, national community and social and ideological antagonisms. Democracies are fragile, they live in conflicts that can overwhelm them: they will remain threatened in the twenty-first century. Furthermore, existing democracies are completely democratic in all their features; there are still limits to democratization within organizations whose efficiency is based on obedience, as in the army. We can ask ourselves, Morin argues, if we could not achieve greater efficiency by appealing to the initiative and responsibility of individuals or groups.

There are processes of democratic regression that tend to expropriate citizens of major political decisions, which tend to atrophy their skills, threaten diversity, and degrade the civic sense. Politics is fragmented into different fields and the possibility of conceiving them together diminishes or disappears. At the same time there is the depoliticization of politics, which self-dissolves in administration, technique (skills/expertise), in economics, in quantifying thinking (polls, statistics). Politics now reduced to shreds, Morin argues, loses the understanding of life, of suffering, of misery, of non-quantifiable needs. All this contributes to democratic regression.

What future, therefore, awaits democracy? The disciplinary developments of the sciences have not only brought, according to Morin, the advantages of the division of labour, but, in addition, the disadvantages of super specialization, compartmentalization, and fragmentation of knowledge, which has become increasingly esoteric and anonymous. In the same way, technical knowledge is reserved for experts, whose competence in a closed sphere is accompanied by incompetence when this same area is impacted by external influences or modified by a new event. In these conditions, citizens lose the right to knowledge. The development of techno-bureaucracy establishes the domain of the expert in all fields relevant to political discussions and decisions.

The fracture between technoscience and citizens, continues Morin, becomes more profound and the duality becomes more acute between those who know and those who do not know. Citizens are excluded from the political spheres, increasingly hoarded by "experts" (p. 60), and the domination of the "new class" (ibid.) effectively prevents the democratization of knowledge. Under conditions like this, the need arises to regenerate democracy and also to regenerate a civic sense, solidarity and responsibility, i.e. realizing what Morin has synthesized in his *Anthropoethics*. Morin's utopian design references Kant's ideas, when he traces the individual loop in order to teach an earth citizenship. The link of the individual with the species has been posited since antiquity. Kant argued that the geographic limits of the Earth impose a principle of universal hospitality.

Our community of planetary destiny must work so that the human species develops in humanity, in common consciousness, and in planetary solidarity of the human race. Humanity, continues Morin, has ceased to be a biological only notion, even though it is inseparable from the biosphere; humanity is rooted in a homeland, the Earth, and the Earth is a homeland in danger. Humanity is a vital reality because it is now, for the very first time, threatened by death. Since the human species continues its adventure under the threat of self-destruction, the imperative has become, according to Morin, to save humanity by realizing it.

A politics of man, a politics of civilization, a reform of thought, *Anthropoethics*, true humanism and the consciousness of Earth-Homeland can only jointly reduce. This is the utopia of Morin, the ignominy in the world. Our ethical and political plan thus requires the democratic strengthening of the individual ↔ society relationship and the strengthening of the individual ↔ species relationship for the realization of humanity, namely the mutual empowerment of the terms of the individual ↔ society ↔ species triad.

From Morin to Illich: School Renovation Within *DemocraCity*

We are living in a dried democracy and culture is a hurried world merchandise.

DemocraCity, utopian city, not thus a physical place, needs rules for its construction. Illich (1973), reminds us that in the capitalist world of exploitation and indifference towards the other, even the school, which must have been the principle of mental freedom, has become the mouthpiece of the current system. The university was once a community of intellectual inquiry, a free zone destined for debate, nowadays it sees students as a key factor in its development: students have become commodities.

Illich argues (1973), that the more education an individual consumes, the greater the wealth of knowledge he acquires and the higher he rises in the hierarchy of the capitalists of knowledge. The teacher, continues Illich, recreates a sacred environment through his figure: it is through his authority, through the aura of respect with which he is held that the sacred enclosure that is the schoolwork. Gluckman (2006), argues that a ritual is a form of behaviour that blinds those who take part, in relation to the gap between the purpose for which they perform the ritual itself and the social consequences that a ritual has.

Schooling is therefore, according to Illich (1973), that ritual that makes us believe that learning can be divided into parts, which can be quantified, and above all that this is a good that must be consumed. Who does not cooperate, continues Illich, who does not consume, is not a modern man. Schooling is a capital investment, but it is also a form of social control, of stratification, it is the creation of a class society [...] with an increasing number of marginalized the more we go up.

For Illich a revolution is possible, a new vision of teaching and learning. We must, continues Illich (1973), recover the responsibility of what we teach or what we learn. Deschooling, according to Illich is a cultural renewal, it is the recovery of the freedom that individuals have to learn and teach beyond institutionalization as a teaching job itself. This mental revolution would lead to what Illich calls conviviality (key element of Degrowth and cornerstone of the Transition Initiatives). It is perhaps Illich that provides the most important brick on which to base *DemocraCity*. He argues that the city of tomorrow is a place in which everyone can be heard, in which no one is obliged to limit the creativity of others, where everyone has the same power to shape the environment which in turn then determines the desires and needs.

The education system could be a real-life laboratory of ideas in the pursuit of a democratic life. School should be the place for learning how to engage in an argumentative debate, to learn the rules necessary for discussion, the awareness of the necessities and procedures for understanding other thoughts, listening, and respecting minority and deviant voices. Learning to understand must play a fundamental role in school learning. It seems impossible at this stage not to mention Michael Billig (2012) who criticizes academia, labelling it's an industry that operates as a chain of publications within the narrow framework of disciplinary orthodoxy, rather than an open laboratory for an eclectic reflection.

This is the very limit of Latouche's degrowth: It remains mired in the exclusive circuit of academia and suffocated by the noise of applause during conferences and symposia. Resilience, therefore, is applicable in many areas and is an indispensable tool to reach the "real utopia", because *homo resaliens* is the one who is able to climb out of a boat overturned after a storm. In conclusion, the parable of the Paraguayan author Eduardo Galeano summarizes the concept of resilience and utopia: "La utopía está en el horizonte. Camino dos pasos, ella se aleja dos pasos y el horizonte se corre diez pasos más allá. ¿Entonces para que sirve la utopía? Para eso, sirve para caminar". (Utopia is on the horizon. I move two steps closer; it moves two steps further away. I walk another ten steps and the horizon runs ten steps further away. As much as I may walk, I'll never reach it. So, what's the point of utopia? The point is this: to keep walking.)

References

Amartya, S. (1982). *Poverty and famines: An essay on entitlement and deprivation*. Oxford University Press, Oxford.

Arendt, H. (2004). *Responsabilità e giudizio, a cura di Jerome Kohn*. Biblioteca Einaudi, Torino.

Beckert, J. (2007). *The great transformation of embeddedness: Karl Polanyi and the new economic sociology* (MPIfG Discussion Paper 07/1).

Berthoud, A. (2005). *Una philosophie de la consommation. Agent e'conomique et sujet moral*. Presses univerisitaires du Septention, Villeneuve d'Ascq.

Bertorello, M., & Corradi, D. (2011). *Capitalismo tossico. Crisi della competizione e modelli alternativi*. Edizioni Alegre.

Billig, M. (2012). Undisciplined beginnings, academic success, and discursive psychology. *British Journal of Social Psychology, 51*, 413–424.

Caillé, A. (1998). *Il terzo paradigma: Antropologia filosofica del dono*. Bollati Boringhieri.

Caillé, A. (2005). *Esquisse d'une analytique du don: Dé-penser l'économique*. La Découverte/MAUSS.

Caillé, A. (2009). *Théorie anti-utilitariste de l'action*. La Découverte/MAUSS, Paris.

Caillé, A., & Chanial, P. (2008). L'amour des autres: Care, compassion, et humanitarisme. *Revue du Mauss, 2*(32), 97–124.

Cardoso Machado, N. M. (2011). Karl Polanyi and the new economic sociology: Notes on the concept of (dis)embeddedness. *RCCS Annual Review, 90*(3), 19–140.
Chanial, P. (2008). *La société vue du don: Manuel de sociologie anti-utilitariste appliquée*. La Découverte/MAUSS.
Dempsey, J., & Robertson, M. (2012). *Ecosystem services: Tensions and developments within neoliberal environmentalism*. Progress in Human Geography.
Fistetti, F. (2008). Is a society of recognition possible? Interview with N. G. Canclini, A. Honneth, and A. Caillé. *Revue Du MAUSS, 32*, 411–432.
Giovannini, E. (2012). *Indagine conoscitiva sull'individuazione di indicatori di misurazione del benessere ulteriori rispetto al PIL*. Audizione del Presidente dell'Istituto nazionale di statistica Enrico Giovannini, Commissione V "Bilancio, tesoro e programmazione" della Camera dei Deputati.
Gluckman, M. (2006). *Politics, law and ritual in tribal society*. Aldine Transaction.
Goldthorpe, J. (1998). Rational action theory for sociology. *The British Journal of Sociology, 49*(2), 167–192.
Illich, I. (1973). *Energy and equity*. Marion Boyars.
Latouche, S. (2007). *La scommessa della decrescita*. Feltrinelli.
Latouche, S. (2009). *Farewell to growth*. Polity Press.
Mauss, M. (2002). *Saggio sul dono. Forma e motivo dello scambio nelle società arcaiche*. Einaudi.
Morin, E. (1999). *Les sept savoirs nécessaires à l'éducation du futur*. UNESCO.
Pallante, M. (2007). *La decrescita felice: La qualità della vita non dipende dal pil*. Editori Riuniti.

CHAPTER 6

Sic Parvis Magna: From Humble Beginnings

Chapter 5 analyses the examples of Transition Initiatives in Italy and Australia, framed as an example of a fragile path towards *DemocraCity*, highlighting not only the seriousness and legitimacy of this new paradigm, but also demonstrating the feasibility of implementation of the concept and as a way of thinking. Chapter five seeks to understand how resilience has affected human communities and the role of resilience in creating better opportunities for realities living in hostile environments or difficult circumstances. What follows is data extracted from conversations, interviews with active members of these virtuous realities, with policy makers and observers of these transitional realities in Australia and Italy.

This research path is an exploratory case study, focusing on Transition Darebin, a small community based in Melbourne. The reasons that influenced the decision to choose this community as an example for this study are threefold: first, Transition Darebin was one of the first initiatives in Melbourne of this kind; second, this community has a link to local politics and politicians (Darebin City Council); and third, and very important, this was also a pragmatic decision as this group was open to be interviewed unlike other existing Transition movements.One of the main activities promoted by TD includes promoting active citizenship through an awareness on climate change, resource depletion and related issues, encouraging the adoption of pro-environmental behaviours, and promoting participation in sustainable community activities. Practical

M. Carcea, *The Road Towards DemocraCity*,
https://doi.org/10.1007/978-981-16-3505-2_6

activities range from permaculture, involving community gardening and social farming, to an environmentally conscious approach to real life, every-day activities such as the ethical purchasing groups, and activities such as lessons and/or workshops with the aim of raising awareness of and thus fighting programmed obsolesce.

The present, (past) and the expectations for the future will be narrated by the members these initiatives as well as their commentary and criticisms of the approaches from Latouche's degrowth and others to the meaning of resilience. All these responses are intertwined with a sense of community and the importance of a new narrative that can provide a valuable tool to reimagine the Degrowth theory.

The New Alliances

Latouche (2007), opened up the panorama with a definition of degrowth, buen vivir, ethical purchasing groups and small cities/neighbourhoods, able to be energetically self-sufficient and ecologically virtuous. One of the limits of degrowth, as pointed out by several scholars is its ineffectiveness. It is not uncommon to find the word "utopia" alongside degrowth.

However, how can we realize a utopia? Perhaps, according to Latouche (2007) it could be achieved by moving from the immaterial to the material state. If Degrowth theory represents the immaterial, i.e. the decolonization of our collective imaginary, the resilience, tool of a virtuous community, is the materialization of degrowth. This journey aims to suggest an alternative narrative, in which three actors of the holistic social-economic area will be combined and integrated and contrasted to the "homo oeconomicus": "homo sustinens" (Siebenhüner, 2000), "homo politicus" (Crouch, 2000), "homoecologicus" (Becker, 2006) in order to find within homo "resaliens" the concretion of the values such as altruism, mutual aid, reciprocity, communication, human values, and responsibility for future generations.

The answer is that the example of Transition Darebin will lead us to the right path for a new narrative, a narrative we call the road to *DemocraCity*. There is also a hierarchical reason behind such a choice: degrowth, as the analysis of the relevant literature has shown, is a discipline based on top-down dynamics (underlined by western social cosmology (as discussed in Chapter 3, p. 77); a Transition Town is instead a community with bottom-up dynamics (bottom-up process). Another underlying reason for undertaking this research is that this topic, despite the urgency and need for countermeasures, is often considered unimportant by social scientists, especially when based on a theoretical assumption that combines Latouche's paradigm, the role of resilience in order to explain the reality based on Buen Vivir, and of course, field observation. The reality is that so far only few studies have been conducted in this field, and those that have been conducted are narrow in scope: they either focus on the economic aspects (mostly based on local and alternative currencies), or on environmental and climate change (energy efficiency, renewable energies).

There are not many academic studies that provide an understanding of the lived experience of Transition Towns, such as the example located in Melbourne, and there are still fewer studies that frame resilience and degrowth as a *unicum* theory, one that will offer the tools for a transition process that leads us to a change of our collective imaginary. The example is why this is an important sample and how it can be the trigger for other examples around the world. As previously discussed, we have long been drawn to the theories of Latouche, Illich, Cochet, Daly and others, however it is the participatory nature of such resilient communities that have fascinated many scholars (Holemans, 2017; Hopkins, 2008; Muraca & Döring, 2018). It was not particular difficult to navigate the website of Transition Towns Australia, and it is possible to find a map of almost all the TTM in Australia within the container website "Earthwise Harmony". Following the dedicated links, the researcher can easily gain an idea of the history of the movement and its main themes such as the peak of oil and climate change.

The first contact with TD consisted of an email sent via the "contact us" link on their website. One of the key figures of TD replied by instructing the study to follow the TD'S Facebook (FB) page in order to allow us to remain up to date with developments of the TD. Scrutinising the FB page for relevant events, for purposes of this study, I decided to attend an event called "The food swap", a small-scale sustainable trade market for fruit and vegetables organized by members of the TD community and open to anyone. As the researcher, I introduced myself as the researcher who had contacted them by email and on Facebook. I explained my project, the main reason for wanting to study themes part of my investigation in the field, my background, my journey to Australia from Italy. At the beginning, very few people made themselves available to me; this was partly due to time constraints on individual member's part; others did not feel "qualified enough to support an academic interview", and some were just wary of me.

I had to attend several events before I was given names and contact details of people considered appropriate to my cause. I came to know a new neighbourhood, and slowly, I began to understand the social structure of this community in which the transition was growing. It was not easy to insert myself into an environment, both physical and social, totally unknown to me and at the same time starting to formulate the most appropriate research techniques found by triangulation of sources: observation, interviews and documents. After a long interview with one of the founders of TD, I was given the opportunity to access their documents via Dropbox. One of the folders contained a chronological report of all events, future project proposals, meeting reports (TD is structured as an INC.), requests for funding from Darebin City Council and further. An important source of information for my research were the self-produced magazines which discussed and synthesized Graeber's analysis, Hopkins concepts, and various themes related to climate change and environmental sustainability.

Transition Darebin attracted me because of its participatory and active nature on the part of the citizens. The initiatives, in fact, start only on the basis of a concrete commitment by someone who wants to participate and contribute to the life of their community, and continue only with a view to collaboration. However, it is not only the ethical dimension, as we will see in the extracts of the interviews, son of the Jonas's imperative of responsibility, very often quoted by Latouche in his work, is a very strong element in the theoretical framework. It is important to stress that the Gift Theory has had a strong influence on such journey.

The "Virtuous" Example of Transition Darebin

Transition Darebin, according to a key figure of Transition Darebin, was founded in 2008 on the basis of the work of Rob Hopkins and his "The Transition Handbook" Thanks to "a fortunate meeting of passionate people", driven by the ethics of responsibility, such reality in 2009 became a reality. This person continues to be a member of TD, "nowadays we are moved, from a linear, almost hierarchical structure that followed the rules (this member of TD was referring to the guideline within The Transition Handbook) to a strategic intent in order to reach a new narrative".

This enthusiastic member explained: "we were four people who then immediately became five (six and so on), however, if you look at our mailing list, we are hundreds, if you look at our events, we can gather a lot of people, unfortunately, in practice, if we talk about active members, we are half a dozen. We are trying to simplify what is called an umbrella organization…" Thus, being an incorporated entity implies, according to this person that members, but also interlocutors, expect a hierarchy and a well-defined structure: "in part that is true, we have roles and key figures, however, in reality things are different…" The roles within the group must serve to connect the points in common, to enable dialogue with other realities and develop new ideas. For instance, the person who holds the role of president must "join the dots" and coordinate the realities in the positive way".

Even being part of the association requires following certain criteria, such as membership where participation is open and free. Transition Darebin, from 2008 to today, have changed. It has moved from how to imagine the transition to how to achieve it. Transition Darebin, perhaps better to say the transition, belongs to anyone who wants to be involved. Involved… It is very interesting because nine people out of ten have different idea of transition, many of them have not heard of Degrowth theory, and some of them barely know their motto "ready and resilient". For instance, the responses by other members of TD to the theoretical structure behind the transition towns were instructive:

"Transition towards what? Degrowth? I honestly do not know what you're talking about… I like to take care of my vegetable garden and exchange its yields to meet people and also because, the vegetables in your house are much better than the products that you can find in the supermarkets". These answers summarized and anticipate the answers the problems well delineated by one of the founding members and key figures of TD:

> I do not care if it's called transition, degrowth or more. On the contrary, if one day, in the near future, a more effective transition movement would jump it will be fine for me… The important thing is to write together a new narrative for Australia, no matter how you want to call it…" This makes us reflect, as evidenced by the interview given by this person: "first you asked me how many we are … we are few, we all have to do something in order to maintain ourselves and our families, and we are not able to change our lifestyle yet….

Cooperation, Resilience, and Local Trade Within Transition Darebin

At a historical moment such as this (and it is not relevant whether we referring to Australia or any other country in the world), characterized by a serious and global economic crisis, listening to people's contributions, ideas (and even sufferings and hopes) is essential to progress the processes of individual and social "resilience", reducing the vulnerability of the population potentially at risk of poverty, marginalization or social exclusion. One of the goals of Transition Darebin, according to some active member with the TD community, is giving back hope and the sense of community. Within the context in which we live, paraphrasing one of the key members, we are called to become a "resilient community", or to think of ourselves as a community that develops intentional actions aimed at strengthening the personal and collective capacity of its members and its institutions to influence the course of a social and economic change. The "sense of community" allows one to share difficulties with others, to recognize each other as peers and to struggle together to negotiate better conditions of life: therefore, it is important to avoid, as one of the TD's cardinal members argues, "priori knowledge". Dialogue, participation, initiative, in a word, the mental shift, recalled by many members of Transition Darebin is the key to achieving a desirable alternative.

However, such mental shift is possible only if we increase the resilience, i.e. as suggested by many members of TD, the structural capacity (in terms of community) to build and improve general resistance, the ability of a system to cope with a series of shocks. One member said, "it is virtuous circle that, thanks to the diversity of the members, works only if we are a group". In the conversation with other members of Transition Darebin, the economic dimension inevitably emerged as a topic and its convergence with the emotional/participatory side regarding the expenses in order to maintain and support the (numerous) activities of the community.

As evidenced by one of the founding members of TD, this reality, i.e. being an incorporated entity, is partly self-financed through membership dues. This person argued: "Sometimes not having money is a positive thing, certainly we avoid waste and corruption…" "Volunteering, reciprocity, mutual aid, and solidarity are the basis of TD, this is the minimum common denominator that has emerged from the testimonies of the members of this initiative". "Money is not the problem", argued an old member of TD, "it is the change of the collective imaginary, we need, again, a mental shift…". "We receive several donations from either members or non-members, however there is no involvement …". Speaking of the economic and voluntary side, I remind the members that in the archives of TD website there was a project related to time banking and the creation of an alternative currency.

"It is true, or initially it was we discussed regarding such matter… However, as you know lots of people are interested into the financial system within TD, but it was complicated: lack of trust, too many expectations, people leaving the community and again, the problem of corruption… Just take as an example Uber or Airbnb: ideas based on a peer-to-peer economic model transformed into a financial business… Nowadays, Airbnb has been an unregistered hotel chain that makes significant profits… The idea of a spare room as a tool to reinforce reciprocity and cohousing no longer exists…Yes, LETSystem is still working, however…".

When one of the pillars of Transition Darebin touched the point of LETS, he reconnected with the question of an alternative currency: "We are looking for a local currency, precursor to a time banking… Looking for a "space", something (similar to a community coworking, social enterprise… Actually, we are in contact with Borderlands, a community action, learning and research centre, located in Hawthorn… We are trying to approach Latrobe University, we need visibility…".

Alternative currencies, sustainable economics, academic support… A very active member of Transition Darebin—described by others as one of the "cornerstones" of the initiative—argued: "we know that are tons of notions and philosophy involved, we are connected with other groups, we are in contact with experts which they have given talks" (referring to the initiative such as *Transition in the Pub*) … "However, nowadays, in reality, we haven't started a real project yet in order to develop our local economy or we don't have any tools to measure our resilient effectiveness… We should learn something from the Christ Church Time Banking model… But it is very time consuming, we have full time jobs, and this is a full-time task if we want this working properly".

"We are living in a disposable society", one of the member of TD argued… "We need to paint a picture of the future. Maybe based on an Eco socialism model", suggests this person. He continues "The truth is, what we are doing nowadays is not new". The *fruit squad*, is an idea—clarified also by other members of TD about rescuing unwanted fruit from the multitude of fruit trees in people's backyards and putting it to good use by giving it to agencies providing food relief service). The *fruit bank*, suggests a member active in its development, is a deposit created to storage unwanted fruit and veggies; and, in addition, the *food swaps*, based once a month in Northcote, Preston/ Thornbury, Reservoir, Fairfield, underlined a member of TD, are not new ideas.

As one of the most active members of TD underlined: "Greeks and Italians in the late 1950s did this all the time…". This is Transition Darebin, a toolbox, a laboratory of ideas, a reality, a community that works together in order to transform the risk into an opportunity. Initiatives such as *The Repair Café* and *Fix it Newlands* (activities established to fight programmed obsolescence of the goods of our disposable society), *Transition in the pub*, and the *Convivial kitchen* are pivotal activities of the transition that is happening now in Darebin.

Members of the transition community work horizontally, trying to create a new local social and economic "fabric", putting people (to be more precise, according to a member of TD, "middle class people") in direct contact and aggregating them to each other.

The Modus Operandi

During my time spent on the field as an observer and scholar of the reality of Transition Darebin, I realized that definitions and concepts such as community, resilience (much appreciated were, according to the members of TD, the model of Kimhi and Shamai (2004) and the one provided by Norris et al. (2008), (for a more in-depth discussion see Chapter 4) and permaculture are very clear. The challenge is creating a new narrative. The goal is "coping with changes, ready for changes". There are also interesting considerations such as "cohousing", "community", "good neighbourhood" and "techno utopia". As already discussed in Chapter 4 it is clear that the concept of cohousing could help the identification of the limits that emerged from my research in field.

Cohousing does not necessarily refer to forms of housing promiscuity. The concept of cohousing does not take the family as a true nucleus of a community structure. The sharing of some spaces and services, can be considered as a tool for socializing and sharing costs and responsibilities. Analysing the historical cohousing roots, Lietaert (2007) makes reference to currents of libertarian and utopian thought. The names often mentioned include Thomas Moore and Charles Fourier and for them the term community is usually carries its clearest and strongest political sense. Lietaert argues that in Danish, the literal meaning of the word for cohousing, *bofaelleskaber*, is "living community". It is no coincidence that these experiences are recorded for the first time during the 1970s, a period known for its renaissance of new thinking and approaches.

Of course, the original "libertarian" background of cohousing, and other influences and new responsibilities are intertwined: the critique of modern life with the interrelated rejection of industrial civilization in favour of a pre-industrial community idea (interesting as one of my interviewees refers to the "indigenous community, as a true community on Australian soil"); the rejection of the urban dimension in favour of a naturalistic or bucolic dimension of life (for instance, as emphasized by one of the founders of TD, "the goal is to change the collective imaginary, to explain and convince people that they do not need a TV, they could enjoy nature outdoors.

"We are here but we are not living here..." argued a Transition Darebin enthusiast, who has recently taken on the ecological side. Without doubt the concept of cohousing offers a tangible form of application of the reduction in cost of living, of reuse and recycling and energy savings, a tangible solution to issues brought into focus by the planetary environmental crisis, brought about by the recent global economic crisis.

Lietaert (2007) emphasized, and the data extracted from the field confirms it, that the phenomenon of loneliness today is characterized according to certain emerging features of life. It is a response to the solitude of urban life, that is a sort of antidote to the structural solitude of those who live in urban centres or, due to aging or for other reasons, can remain alone and isolated (interesting, for instance, the words pronounced by a member of TD and confirmed by other members during my research on field: "between general meetings and TD events, I see member of our group more than once a month, and I'm sure if you think about it and it's more than you see your friends …".

Moreover, in the general context of the crisis of welfare state models, cohousing, using the lesson provided by Lietart (2007), can serve as a form of reaction that, starting from the bottom, proposes to give concrete answers, and with a certain autonomy with respect to local administrations, to the needs of children and the elderly. The sharing of burdens between co-residents would thus compensate for the structural weaknesses of public administrations. The lowest common denominator of this initiative (regardless of the fact that only a small percentage of my interviewees cohabit in the same house) is shared planning. Transition Darebin (but we can apply such quality on every single transition initiative) has a vision.

Paraphrasing Diane Leafe Christian (2003), a vision is not a simple collection of words (for instance, the TD's motto Ready and Resilient), it is a ray of energy that drives a group. The vision must therefore be formalized in a declaration, continues Diane Leafe Christian: a structured vision statement offers a clear, concise, and convincing expression of the group's vision and mission (naturally TD has is constitution as well, available on their website). Diane Leafe Christian argues that the missions and the cardinal points of the group must have short names that remain imprinted in the minds of people (and, of course, in the minds of the members). It is therefore not a coincidence that activities, initiatives of TD, have names like *Repair Cafe*, *Food swaps*, *Convivial kitchen* etc.

As mentioned above, although TD is not a community based on cohousing, it embraces the philosophy or features of weak communities. As Lietaert (2007) argues because the general and common cultural horizon that animates such models, albeit with many nuances, presents itself as fundamentally ambiguous in its terms from the original formulations and realizations. Secondly, because such projects seem to be constitutively lacking, and today more than ever, a broader vision, of a high political framework and endowed with a broader project. Another limit, according to our data, would seem that the communities dedicated to the transition initiatives would favour the tendency towards self-sufficiency, however this factor transforms them into isolated communities.

The transition initiatives favour the establishment of small communities that, with motivations more or less shared at the from the beginning, and without aspirations for broader and shared critical reflection which can therefore disregard the more general global situation. That seems to be a reminder of the Voltaire's lesson "one must cultivate one's own garden". According to the data, the initiatives of TD seem to work well for those belonging to the middle classes who show a certain openness of mind in cultural preferences and who have the necessary economic well-being and the right political reputation to enter into bargaining with local or state administrations.

But the data analysis shows that political orientation is not an important element for community cohesion. However, we believe that the Transition Initiatives can realize some soft transition for a weak social utopia (some members and enthusiasts mentioned an "eco socialism") whose centre appears to be, in a certain sense, the individual and not the community. There is no doubt that this case study can be coherent with some proposals for the "concrete utopia" of Latouche's "serene and convivial degrowth". According to the data it seems, that such a reality can fall into one of his "8 Rs" i.e. the R of "Relocalize", which not only concerns the production and consumption of goods, but the very dimension of urban areas and their economic and political autonomy.

Transition Darebin in this sense could be understood as the affirmation of a new way of conceiving the "local" which, according to Latouche (2009), is not a closed microcosm, but a node in a network of virtuous and supportive transversal relationships, aimed at experimenting with practices to strengthen democracy that allow resistance to the liberalist domination. Ironically, none of the members of TD carried out in the field, neither the members of the community nor members of the political class of Darebin City Council had ever heard of concrete utopia, Serge Latouche and his Degrowth paradigm, before.

We used the word "ironically" on purpose because the members of this transition initiative are very frightened by the word degrowth. One of the key figures behind the Fruit squad rejected the term under examination and questioned by proposing the term "regrowth" i.e. "a rediscovery of values and practices, because what we are doing is nothing new..." and at the same time apply some cardinal points of thought of such paradigm that could be a cornerstone of *DemocraCity*.

As already discussed, economic well-being does not coincide with the wellness of the individual; contemporary concepts of wellness focus on the well-have (Matthey, 2010). The transition movement does not see Degrowth as a return to post-atomic society (the *Mad Max scenario*, recalled by one of the cornerstones of Transition Darebin), but it hopes for what I call the rise of *homo resaliens* (from the Latin *resalio*—no longer *resiliere*—that is the ability to arise, again, from a boat overturned after a storm). Degrowth is a movement that fights an ideological battle starting from scholarly, political, and institutional environments. It tends to follow dynamics from the top to the bottom (*top-down*) and the action of citizens follows the sensitization of decision-making and power bodies.

Degrowth is a movement that fights an ideological battle starting from all sectors of our complex society. It tends to follow dynamics from the top to the bottom (*top-down*) and the action of citizens follows the sensitization of decision-making and power bodies. The challenge is fundamentally intellectual (the decolonization of our collective imaginary), even if something has already changed in the direction of the Degrowth theory. The problem of the "laboratories of ideas" based on the theory promoted by Latouche is that they follow the vision of the French economist: they have been reduced to slogans and provocative actions. This reduction, combined with the multidisciplinary nature of the subject, has meant that Degrowth theory is perceived exclusively as an academic dissertation.

According to the data, the transition is born from a *snowball effect*. The transition explanation, which Transition Darebin call "*Transition in the pub*", indicates the main topics and challenges of the TTM, i.e. climate change and oil peak, framed as an interconnected problem. Here the difference between a manual and a treatize is highlighted. The handbook written by Hopkins is a valuable source that guides the community step by step, both from the point of view of the ideological and conceptual structure and from the practical organization: from the strengthening of bonds of trust within the community, to forums intended as a laboratory of ideas, to assemblies open to new participants with the intention of attracting external people through practical demonstrations.

The usefulness of social networks and web portals in the diffusion of these communities is undeniable, but in truth, it is due to the desire for active citizenship and to exercise that right of inclusive democracy by its the members, that both degrowth and transition, can exist. Transition takes its inspiration, as we will see from the data, from Latouche's virtuous circle of 8 Rs, but in addition, from Norris et al. (2008) 3 Rs for resilient communities and the three main characteristics identified by Kimhi and Shamai (2004) i.e. propensity for resistance, recovery trend, and propensity for creativity. In a certain sense, the transition completes and exceeds the theoretical model of Latouche's degrowth.

The Transition Town Movement mentioned by Latouche (2013) emerged through the work and experiments of Professor Rob Hopkins in 2004 in his The Transition Handbook: *from oil dependency to local resilience* (2008). The TTM is an ecological, political movement that has been growing since 2004. It is about communities stepping up to address the big challenges they face. The starting point is local. By coming together, members of the community are able to crowd-source solutions.

They seek to nurture a caring culture, one focused on supporting each other, both as groups or as wider communities (transitionnetwork.org). The TTM is based on a pragmatic approach that underlines the importance of practical activities: renewable energy facilities, local food provision, local currencies, free trade, sustainable market), which offer concrete outcomes within the community and encourage citizens to get involved in this process. However, a further difference is the political orientation: Latouche's position has a political basis somewhat to the left, Illich's small left masonry (Latouche, 2013), while in the TTM a political framework exists but is not necessarily relevant at all times as the approach is predominantly practical.

This difference becomes obvious when comparing the titles of the reference texts of these two alternatives: *Handbook* vs a *Petit traité*, i.e. the practical approach versus the intellectual one. The TTM is the practical answer of a movement that is spreading, not only in Melbourne (Australia) but globally. It is based on a dynamic from below (bottom-p) and involves participatory action by citizens who have as a goal to achieve a change that involves the community in its entirety and to reinforce the principles of reciprocity. According to Hopkins (2008) the TTM is framed as a genuine example of active citizenship, working through sub-politics and drawing on citizens who do not consider themselves activists.

One of the mainstays of TTM is the willingness to enter into relationships with local policy makers in order to create a resilient and sustainable community. This willingness has led to two specific criticisms of the movement and its members: TTM show an avoidance of political confrontation and have adopted a cynical and detached view of politics. One of my question is how long TTM can survive and thrive without any political involvement.

Citing Steffen (2009): "all over the world, groups of people with graduate degrees, affluence, decades of work experience, varieties of advanced training and technological capacities beyond the imagining of our great-grandparents are coming together, looking into the face of apocalypse… and deciding to start a seed exchange or a kid clothing swap".

That risk of confusion and vacuity stressed by Van den Bergh applies to Degrowth theory as well as to the theoretical basis of TTM. According to Chatterton and Cutler (2008), another criticism aimed at the TTM is that avoiding confrontation in the area of local politics is a risk that jeopardizes the integrity of the community.

The Example of Fiorivano le viole

"Sfiorivano le viole"/"The violets faded". This is the title of the famous song by Rino Gaetano. In Perugia instead in via della viola they flourish again, thanks to the cultural association "Fiorivano le viole", where a group of boys and girls who care about Perugia. This group of people tried to enhance this street a few steps from the centre, which was by now in decay, trying to create a community open to participation and also focusing on the theme of integration. This, in my opinion, is a story worth telling.

To date, if you walk down the street and look around you find yourself catapulted into a world of colour and joy, where the paintings of various artists illuminate the walls of one of the streets of Perugia which until recently was one of the "darkest" of the Old Town.

It all begins with "Braccia Rubate", an art workshop founded by Francesco Capponi and Chiara Dionigi in one of the many abandoned spaces in the Via della Viola district, four years ago these spaces were recovered by the association and through a concession of the owners, assigned to various artists to make creative workshops.

From "Braccia Rubate" art workshop it has become "Creative workshop the violets bloomed", a place dedicated to meetings, conferences, workshops but also "aperitifs and improvised meetings". Emblematic are the words of their manifesto that are also found on the walls of the street: *We are a cell of creative resistance against the cultural and civil degradation that grips our streets, our places, our feelings. We believe in a union as a possibility of rebirth, as a creative and necessary instance of being, we believe in the beauty of art, in the sinuosity of forms, in the depth of music, in the kindness of plants, in the love between men …*

Observing the members of the association, their desire to do, but above all to carry on the project they started, emerges, there have been many initiatives on their part, such as an Italian course for foreigners to try to break down those barriers which in the first place the language places you, but also markets and concerts on holidays, which have given new life to this previously forgotten street. One of the most important events organized by the association is Alchemika, the festival of street artists now in its third edition. An international event that brings street theatre, circus art, poetry, and music to Perugia; thanks to the various artists you can immerse yourself in a magical, almost dreamlike atmosphere in the streets of Perugia.

Exhibitions and workshops continue and on Sunday 6 November, thanks to the FRA project—analogue resistance front—an exhibition was held where the photos selected for the contest held during Alchemika 2016 were displayed and printed the same day in a workshop on cyanotype, also scheduled for December 3 there will be an ethnic market, a good opportunity to create a new meeting point for the inhabitants of the city.

The events will not be as frequent as during the most intense moment of work of the association, but nothing can be reproached to these guys who with little money and a lot of creativity have tried to change things, aiming for a real rebirth of the way that by now full colour becomes a characteristic of the historic centre. There has been a change, the workshops of the artisans have remained, some places have opened and for those who already existed, things are better.

A quality cinema, the Post Modernissimo which reopens and certainly contributes to giving new vitality to the neighbourhood. "Fiorivano le viole" is an association that has done a lot, which has focused on art, culture, and integration, trying to break down the distance between people, bringing them closer, making them more human, an association that should be encouraged and supported, which has transformed a reality making it more liveable and socially connected: the rebirth of via della viola.

Leaving Competitiveness

The relocation of companies and public functions, the creation of real peripheral commercial areas, the widespread university model: Perugia is a city that has changed rapidly and contradicting itself over the decades. The historic centre in particular, from a place of shared identity par excellence, slowly transforms itself into a degraded environment: the sense of abandonment raises the threshold of indifference, the level of insecurity increases, and episodes of crime multiply. A sort of self-expulsion disintegrates over the years, the old Perugian society and the historic districts are left to students and immigrants who have recently arrived in the city.

Via della Viola, does not stand out from the rest of the agropolis: one after the other the commercial activities close their doors; restless residents retreat into their homes; drug dealing and degradation are the masters. And yet, there is a ferment in the air: a creative and combative minority aims to reconquer the disputed place and the "call to the arts" arrives.

In Italy, precisely in Perugia, three years after the founding of TD in Melbourne, paraphrasing the word of a member of the association, Fiorivano le Viole was born from the common desire to put an end to a situation of degradation that has lasted too long.

According to the cornerstone of Fiorivano le viole, this cultural movement moved in a concrete and determined way. The core members behind the association asked the owners to grant for the cause the use of their vacant spaces to be transformed into art workshops. At the beginning the collaborators were 14: artists flocked from all over. Via della viola began to hope again, to produce culture, art, above all sociality. When the neighbourhood was closed, dark, abandoned, it began to live a new existence, transforming itself into a permanent art installation, a garrison of poetry and love, a daily destination for tourists and onlookers.

At the beginning, paraphrasing the words of a key member of FLV, the effort was focused on advertising the association, trying to attract an outside audience. Hence the idea of cultural events to be allocated to the entire community. In 2013 we started with the cinema and theatre review; from Donati & Olesen, to the company "Artisti sul Mondo", up to Giampiero Frondini who staged the famous show by Dario Fo, "Lo santo jullare Francesco": there were several local artists who collaborated for free on the project.

The key person behind FLV underlined that the economic return for the district was immediate: while events open to the public were held almost on a monthly basis, young people returned to circulate on the streets and businesses to reopen. The transformation process therefore had beneficial effects on the entire community but the starting point, the president is keen to emphasize, was and remains a cultural necessity. Investing in culture, according to the president of FLV, was the key for the foundations for subsequent economic development, culture and art have been the leverage for everything else.

The common denominator between TD and FLV is the relationship with local politics. A member of FLV underlined that as a cultural association, FLV is outside the political logic; everything was achieved in complete autonomy. However, this estimate member of the association stressed that the relationship with the administrations that have followed over time has always been positive. All this has led Via della viola to be a permanent outdoor exhibition. It must be remembered that many of the initiatives are supported by regional and European funds. Although everyone collaborates totally free of charge in the whole project, the funds allow the association to put new ideas into practice. As we already mentioned in the previous paragraph, Alchemika, the street theatre (but also the cirque nouveau festival, as have underlined a key person behind the association) that takes place every year, is the most obvious symbol. The event that has been taking place since 2013 has now become a fixed and long-awaited event and the proof of this is the great participation it gathers every year.

A Modern Myth

In the first chapter of the anthology Fleeing Vesuvius: Overcoming the Risks of Economic and Environmental Collapse, titled "On the Cusp of Collapse: Complexity, Energy and the Globalized Economy", David Korowicz showed how fragile the current world system has become. Relatively minor events, for instance the eruption of a volcano in Iceland, or the collapse of Lehman Brothers in 2006, had major effects around the world. He explained this as the consequence of interdependence and lock-ins in the world system, providing little freedom to respond to perturbations.

The Covid-19 pandemic is a bigger perturbation than the events he described, although it is less severe than it could have been. That such a pandemic could occur was not unexpected. It had been predicted by Bill Gates almost exactly as it happened only a few years ago. It could have been far worse, with a death rate of MERS (about 30%) instead of 1%, and given the frequency of new epidemics, it is likely that there will be another pandemic with a much higher mortality rate conjoined with the contagious qualities of COVID-19. The current pandemic has highlighted how the conditions for such a pandemic have been created, which itself could have a major effect.

All this did not stop FLV and, as reminded us by many volunteers, it underlined the need to make the presence of the association stronger, by activating various solidarity initiatives: from the solidarity basket, a suspended shopping service made possible thanks to the collaboration of the Coop (an Italian supermarket network) of the city centre, to that of the solidarity mail, which allows the inhabitants of the neighbourhood to leave a specific request for help to the association. In addition, another project is active: the suspended drug service.

The project carried out by almost 150 volunteers of the association of Fiorivano le viole makes it clear how the spontaneous aggregation of a community that is rediscovered around a common project, allows to creatively overcome problematic situations and how access to resources cultural can be facilitated or hindered by the social dynamics that are established.

If degrowth "scares" and if a growth means absence of growth, perhaps the terms "regrowth" and "rediscovery", taken from the study in the field may remind us of the lesson of Hervè Kempf (2010): the author argued that we do not invent the new world, it already exists, not exploited, as a soil ready to germinate and produce a golden harvest.

However, in order for there to be a rich production of a golden harvest, it is essential that the seeds all grow in a coordinated way. In the transition to *DemocraCity*, we must start from concepts such as Natural Capitalism (Hawken et al., 2001), and a resilient industrial design, supported by the concept of a Product Service System (PSS): an industrial system based on a radical increase in resource productivity, industries based on closed-loop biological models (zero waste), moving from the sale of goods to the provision of services, and reinvesting in natural capital.

It is necessary to design a political and an academic vision that combines concepts such as Cradle to Cradle, Biomimicry and Blue Economy, in order to build the foundations of this qualitative regrowth. The propensity the drive to creativity, set out by Kimhi and Shamai (2004), within the resilient communities, pushes towards tensions that can favour forms of social innovation. In such a context, homo resaliens becomes a social actor that thanks to the cultural and operational tools available can feed and support such regrowth processes in which all of us, experts and non-experts, should be involved.

The challenge that awaits us is therefore to overcome the impasse. We are living in a world where two realities coexist in a conflicting way: the old world without limits and another that recognizes these limits and experiments and applies the lesson of Edgar Morin (1985): crisis in Greek (krino) means separate, i.e. choose the good part over the bad part (diagnosis and prognosis; problem solving and decision-making). Nakamura and Kijima (2009) remind us how the same ideogram in Chinese is used for words such as "danger" and "opportunity". It is time to pursue new ways to transform the old limits into opportunities. The new narrative must be the transition to a different model of development in which homo resaliens must excel in problem solving and decision and sense making.

Pauli (2014) argued that the aims of the Blue Economy (a new type of economy that is based primarily on natural systems) are characterized by entrepreneurial opportunities inspired by ecosystems. Pauli points out that what we see developing before our eyes goes beyond the genius of each species: the total is worth more than the sum of the parts. A world that hosts marvellous ecosystems, Pauli stressed, that is capable of letting fall nutrients and energy cascading on multiple levels, without end, is a world that can accept the challenge of solving poverty and misery, inequality and waste. Homo resaliens has to accept a different economic vision.

The same Circular Economy (not to be confused with Cradle to Cradle) has timidly become part of the recent European policies. According to the Ellen MacArthur Foundation (2017), Circular Economy is a generic term to define an economy designed for to regenerate on its own. In a circular economy the flows of materials are of two types: the biological ones, capable of being reintegrated into the biosphere, and the technical ones, destined to be revaluated without entering the biosphere. The circular economy is therefore, according to Economic Growth Potential of More Circular Economy (2017), is a system in which all the activities, starting from extraction and production, are organized in a way of production that someone's waste becomes resources for someone else. From a production point of view, the system is moving from a linear concept to a circular type model. However, this is not enough: linear chain "take-make-dispose" models can be improved, optimized, they can become efficient, but as long as waste remains, pollutants and waste, industrial production and consumption will continue to discharge externalities. It seems clear to us that linear production, which requires a constant increase in resources and productivity, is faced with the limits of the biosphere.

The circular economy defines, unlike the linear system, is a processively regenerative industrial system in a continuous loop. From the concept of end of life, the infamous programmed obsolescence, we move on to the requalification and reconstruction, we move towards the use of renewable energy. The goal is to eradicate waste through an innovative design of materials, products, systems, and a new economic model. The problem, for which sustainable development models are still defined as toxic by many scholars and growth objectors is that circular system economy foresees two material flows: a biological cycle and a technical one. The problem lies in the part of the technical process where the materials defined as "technological nutrients" can be of mineral or fossil origin. Transforming the flow of materials to build a perfect circle and is the challenge that awaits us despite the launch of the concept by the German chemist Michael Braungart in the 1980s.

This type of challenge must be overcome by homo resaliens, the one who is able to perform that mental shift, who no longer sees himself as a devotee of consumption, but as a social actor who uses a good without possessing it. Fundamental keys for the prosperity of such an economy will be rental and leasing. McDonough and Braungart (2002) argued that switching from the concept of eco-efficiency to the one of effectiveness is a necessary achievement in order to improve a desirable production system: we should remember here that the concept of efficiency in economics aims to use clean technologies, seeking only to minimize the volume, speed and toxicity of material flows. Only some materials are recycled, others are designed following the concept of *end-of-pipe*.

According to this matter, Cradle to Cradle economy proposes the of eco-efficacy solution in order to re-establish a synergistic relationship between natural and industrial systems, thus converting anthropic and economic processes to biological processes. Therefore, the holistic view emphasized by Gunter Pauli (2014) in opposition to the reductio ad unum through the work of the homo oeconomicus and his anthropocentrism is interesting. The author, pioneer of the Blue Economy, in order to explain the relation between man, economics and ecosystem argued that the word invention is *quasi* looting what happens naturally within the ecosystems.

Rethinking Degrowth: Disappointment or Need?

As described, the analysis of the relevant literature dedicated to degrowth demonstrated that degrowth theorists consider it necessary question the dominant world view through a process of deconstruction of the imaginary. Starting from this idea, the same scholars criticize the whole structure of knowledge as it is presented today, of the techniques, methodologies and dominant scientific paradigms. Erich Fromm (1977) used the word *Ent-täuschung* referring to the idea that all (true) knowledge begins with the demolition of illusions. What these scholars question, including my mentor in this journey, the Australian philosopher Arran Gare (2000), is the irrational faith in Western science and the belief in the omnipotence of technology, thus expressing a firm opposition to Promethean sciences, whose only end is dominating nature. Reconceptualizing disciplines, teachings, and the entire scientific universe means considering the ambiguity of knowledge in his being unique and at the same time manifold.

This is the challenge the paradigm of degrowth faces, ignored by my sample of respondents, placing itself within the vein of critical thinking that the humanistic sciences formulate on current trends in the contemporary world. The aforementioned Edgar Morin (1999) argues that human beings are at the same time physical, biological, cultural, social, and historical. This complex unity of human nature according to Morin, completely disintegrates in teaching, because of a division into various disciplines. Today, according to the French philosopher, it is impossible to learn what it means to be human, while every individual independent of location, should have a knowledge and awareness of both the complex character of his identity and the identity he has in common with all other humans. Morin also considers it necessary to recognize the unity and complexity of the human being by synthesizing and organizing the knowledge dispersed in the natural sciences, in the human sciences, in literature and in philosophy.

Latouche affirms (2012) that the time is now ripe if we want to redirect scientific and technical research on the basis of new aspirations. The productive and reductionist approach determines the choices and the dominant orientations in the research fields. The importance of dimensions linked to visibility, to spectacularizing and to the laws of the market to the detriment of effectively utilizing research for social purposes drives a large part of the system of knowledge. Fromm argued (1977) that ours is a society in which the market dominates, in which appearance is everything and image acquires enormous importance (and view is confirmed by one of the TD members who felt that we live in a "disposable society").

This journey proves that within the modern and globalized academy, the production of knowledge does not leave any space for the construction of knowledge nor the creation of knowledge. The colonization of our collective imaginary, and therefore of the system of meanings, beliefs, and attitudes, which has flooded our lives for centuries, is present and is also imposed in the academic sphere, influencing research practices and achievements. From this perspective, the contribution of degrowth is a useful tool to rethink thinking and produce new knowledge.

The need to rethink our way of living, according to von Glasersfeld et al. (1994) and being in the world therefore implies a re-orientation of research, techniques and scientific practices in an interdisciplinary and transdisciplinary universe, complex and varied, reconstructing epistemologies and systems of knowledge aimed at the collective rather than private, putting at the centre of its interest not the product and the profit, nor the self-referential and dominant man, but the man in nature. The autopoiesis, (the property of a living system capable of reproducing and maintaining itself) using the lesson of Maturana and Varela (1987), risks becoming confused with the self-referentiality of man's power over everything, of man seen not as belonging to a complex ecological system and in which he is a guest but of the man who claims to control, exploit and determine the planet at will without any consideration of the global ecosystem.

Fromm argued (1977) that the original unity with nature has been completely lost and, in order not to feel completely alienated, it is necessary to find a new unity, both with our fellow beings and with nature. If we consider social justice and the principle of responsibility as parts of the dimensions that should guide social and political action aimed at well-being, it seems essential to redefine this concept with respect to the use of equally distributed resources, considering in this redefinition a centrality of man in nature as not all encompassing, and the temporal aspect of generational inheritance that accompanies the human species.

According to Latouche (2012) a greater attention to ecology and future generations allows us to more fully appreciate the complexity and to re-establish the terms of social justice and community rights and duties by approaching the world view expressed in the concept of frugal abundance. As repeatedly emphasized by the data collected, the deconstruction of our collective imaginary requires a radical change in needs, choices and expectations by everyone. The transition towards *DemocraCity* needs joint effort from History, Anthropology, Economics, Political Sciences, and Technology Studies.

Whose Future Is It?

Change can take place to the extent that it starts from ourselves; however, it must absolutely overcome the dynamics of bottom-up/top-down that characterize the paradigms of degrowth and transition. It becomes crucial to overcome the psychological resistance and the underlying definitions of anti-group, anti-community and weak community. Our freedom from oppression by the dominant paradigm can only take place when we consider the whole of humanity as oppressed by itself. It is necessary to establish a global terrestrial identity, as Morin (2002) has argued, which unites in differences and allows overcoming of the only human limit that is worth overcoming: the illusion of having no limits and being omnipotent. The words of the French philosopher bring me back to the words of a key member of Transition Darebin: “we have to contrast the Mad Max scenario… A new narrative is fundamental…” (Carcea, 2019).

If the new narrative, the new regrowth, created to avoid such a scenario, will be the fusion of a natural and a technological intelligence? Technological innovation in a society based on infinite growth is oriented towards increasing production capacity and consumption, a trend that has done nothing but increase environmental degradation and put at risk the survival of mankind. Following the path of the inevitable catastrophe can prove counterproductive and harmful, despite the temptation being strong due to its simplicity. It seems crucial that the reflection goes beyond and touches the concreteness of a real revolution led by the “regrowth” of all those skills and technologies whose objectives become those of efficiency in the use of resources such as materials and energy, fighting programmed obsolescence of goods, and reusing and recycling the materials of which they are composed.

A rediscovery (or a regrowth) led by “happy degrowth” (Pallante, 2007) of the ecological footprint of an economy that must introduce to our current society a new and unknown model of scarcity. Paraphrasing Hawken et al. (2007), at the beginning of the industrial revolution, the workforce was relatively smaller (about a tenth of the current one), while global stocks of natural capital were plentiful and largely unexploited. Nowadays, the situation has been reversed dramatically.

After two centuries of productivity growth, the exploitation dismemberment of resources at the pure costs of their extraction, the exploitation of living systems as if they were an infinite good and continually renewed, has turned people into an abundant resource, while nature has become scarce in a disturbing way. Hawken et al. argued (2007) that if we want to apply the same economic logic of the industrial revolution to this new model of scarcity with the aim of future prosperity, it is necessary that society make resources much more productive, drawing on everything from the planet (energy, water, matter) at four, ten or even a hundred times greater benefit.

It is important, indeed fundamental, as Pallante (2007) reminds us to start the change process of our collective imaginary by overthrowing and reconceptualizing the hierarchy that regulates our economic system. In the normative and within governmental process that will regulate a society that embraces the principles of degrowth theory, human capital and natural capital can no longer be slaves to fixed and financial capital. Another pillar will be the role of (smart) technology. Here we are not referring to the unshakable faith, to the exaggerated optimism that the objectors of degrowth nourish towards technological progress (impossible to forget how, in 1974, Robert Solow, in his belief in the power of technology, strongly rejected the worrying report accomplished by the Club of Rome regarding world ecological sustainability).

Planetary Understanding, Ethics, and Culture Within *DemocraCity*

It is not possible to deny the importance of technological processes and in particular of those innovative processes, both at a production and social level, that forge new and unexplored paths outside the dominant logic. Rather it should be stressed that natural capital, defined by the sum of the ecological systems that sustain life, cannot be produced or even replaced by human activity. Interesting, thus, in his context is the lesson by the Japanese philosopher Tomunubo Imanichi. In his *eco-ethica* Imanichi explores (2009) the issues arising from the changed habitat and our changing moral consciousness and presents an ethics that transcends interpersonal ethics so as to also encompass companies and government.

The Japanese author (2009) advocates learning from nature and argues that the human race has an ethical responsibility towards nature and matters, including life itself. By demonstrating that virtues were created as necessary in the past, he raises the possibility of creating new virtues to meet contemporary needs. An optimism replaced therefore by a prudent technological scepticism (term very close to the growth objectors) that raises serious doubts about the possibility and desirability of overcoming natural limits through technological advances.

Paraphrasing Costanza et al. (1997), an opposing line of thought assumes that technology will not be able to circumvent fundamental energy and resource constraints and that eventually material economic growth will stop. It has usually been ecologists and other life scientists that take this point of view (notable exceptions among economists are

J. S. Mill, Georgescu-Roegen, Boulding, and Daly), largely because they study natural systems that invariably do stop growing when they reach fundamental resource constraints.

A healthy ecosystem, according to Costanza is one that maintains a stable level of development. Unlimited growth eventually becomes cancerous, not healthy, in this view. It is necessary to recognize the limits and it is fundamental that old technologies must be evaluated not in the light of their private economic productivity, but by virtue of their social costs. Today more than ever, at a distance of 10 years from the disengagement of the "semantic bomb" by Latouche, we must ask ourselves what are the technologies that individuals really need, technologies that really contribute to well-being rather than well-having.

Indeed, the basis for regrowth had already been identified by Commoner in 1986. He argued that the current technologies implied in the production system must be redesigned to conform as much as possible to ecological needs. This means, according to Commoner (1974) that new investments, in terms of both agricultural and industrial production and transport, should be driven above all by ecological considerations, so that the global investment framework should be subsumed under the control of ecological imperatives rather than economic conventional imperatives. As Kapp has argued (1991), economic expansion was generated by investment decisions carried out without prior assessment of economic, social and ecological consequences.

While profits have been internalized and provided the economic justification and resources for further expansion in the same direction, the problems, according to Kapp, of environmental degradation and pollution control through appropriate instruments have been ignored or. (At best) their solution has been postponed to the future. These are the reasons why concepts such as productivity of resources, eco-efficiency as Bertorello and Corradi (2011) remind us, are toxic words.

However, denial that our well-having is increased by technological innovation is hypocritical, A new paradigm must shift from a manipulative knowledge of nature, which selects and simplifies the systems under study, to a knowledge aimed at deepening the complex interweaving of connections between the different systems. The crude simplification of natural phenomena to mechanical phenomena, must be substituted by an analysis of the complexity of the systems, interacting with each other.

This natural epistemology is a necessary premise to start the path towards *DemocraCity*, in which human activities do not reduce all tangible and intangible goods to merchandise but uses in such a way that they can be inserted into complex and delicate dynamic equilibria, present in the natural environment, without destroying them, without transforming resources into waste, without reducing the biodiversity of living organisms. At the moment, scientific research is not free, because it depends on lobbies, who request it and who finance it, mostly centres of economic and financial power, which do not involve citizens, and often do not even employ scientists; also, many scientists and academics do not have the monitoring of research financed by private individuals.

Similarly, the technological applications of such research, covered by patents or industrial secrets, have social, environmental, and health impacts for the entire population, which is excluded from the decisions and often kept in the dark about possible repercussions. A new scientific paradigm also requires popular patronage and collective decisions on technological applications, especially as a function of the social needs of communities, respecting the environment and biogeochemical cycles. The knowledge that emerges from such research constitutes a common good that, starting with the community that has enabled such knowledge creation, must be, as a "common good" (like all the knowledge), freely available to the entire human community, without barriers and without commodification processes.

The necessary realization of sustainable and really useful objects and technologies, which do not have the purpose of profit, but a fair income for those who participate in their production, requires a participative democratic process, which involves the whole community. A degrowth society will not be able to reach its goal without scientific research and technological applications. However, it will not be able to delegate such research to researchers, technocrats or centres of economic power and above all, it will not be able to entrust its present and its future to scientific evaluations only, with the awareness that science and technology cannot transcend natural limits, such as the depletion of resources, but only respond to social questions in a probabilistic and non-deterministic way, as the current mechanistic paradigm holds. A concrete technological reflection cannot, however, be exhausted on a purely theoretical level, despite the previous guidelines, and especially the Herman Daly's laws of sustainability of, constitute important references that should not be disregarded.

Smart Technology and the Regrowth of an Economic System

Pauli (2014) emphasized this bond referring to the first light bulb created by Edison: Pauli argued that Edison used a bamboo filament, a material naturally rich in iron. According to Pauli, Edison did not invent electricity: cells, had been using it for millions of years. It is necessary, according to Pauli, to transform problems into opportunities and to base the economy on the inspiration taken from natural systems. He stresses that once oxygen was also toxic, yet it was the prerequisite of life on earth. For instance, a step forward in order to respond to the CO_2 problem could be targeted use of algae that need a process of photosynthesis and therefore CO_2, water, nutrients, and sunlight.

Let us take another step forward and stretch our perspectives: a single hummingbird uses the nectar of 1,000 flowers in a day, goes up to 56 kms per hour, migrates about 3,218.7 km per year, and to make a journey of 965 km, and uses 2.1 g of nectar as fuel. Or perhaps let us consider the example provided by Janine Benyus (1997), regarding the classic and conventional process of calcareous material transformation in construction material: from cutting into blocks, shredding and firing at 1,500° C, and what a hen does, which in a few hours produces a much more resistant eggshell; or again, between Kevlar, an oil-derived material that can block bullets, and the spider's web, just as strong and much more resistant. Biomimicry and Cradle to Cradle economy must work together synergistically in the construction of the aforementioned narrative. In 1958, Jake Steele (an aeronautical engineer) coined the term "bionics" describing the science of systems based on the workings of natural systems.

The observation of nature and its phenomena is a promising vision for a newfound alternative. Homo resaliens act as a bridge thanks to his multi and interdisciplinary contribution and his heterogeneous skills. Biomimicry (or Biomimetics) is a science that studies the models of nature to imitate or take inspiration from these to solve human problems. Benyus 1997, biologist founder of Biomimicry 3.8 argued that we should consider nature as an "inventor": with about 3.8 billion years of innovation behind it, which man should observe to find out what the s secret r of so much efficiency and zero emissions is. Biomimetics uses an interdisciplinary approach combining disconnected and fragmented worlds: nature and technology, biology and innovation, life and design. In its most practical form, biomimetics is a way of looking for sustainable solutions by borrowing life models, chemical recipes, and ecosystem strategies.

We must remember Munari's lesson (1997) so that it does not end in oblivion "Orange—The subject is a series of modulated, segmented containers, arranged in a circle around a vertical axis with their straight side leaning against the axis and the round side pointing outward to appear spherical. All these segments are packages with a well-characterized material and colour: fairly hard on the outside and soft on the inside, to protect the outside and the containers… Every container is made of plastic sheeting, which can contain the juice and still be easily moved.

A very weak glue keeps all these containers together, so it can be taken apart at any time. The packaging—as is common nowadays—need not be returned to the manufacturer but can be thrown away… So, this orange is an almost perfect object with absolute consistency between shape, functionality, and consumption. Even the colour is right: blue would be a mistake" (pp. 136–137).

A further piece that completes the mosaic that depicts the technologies behind such regrowth process is overturning the logic that underlies the operation of the economic system: the looting of natural capital. Access to resources plays an increasingly problematic role in the modern functioning of the market, in consideration of the fact that when it comes to the flow of resources and services from the environment, the criticality is not just about raw materials or fossil fuels such as oil, but it is also extendable to the per capita availability of water, arable land, and even fish.

While this study was proceeding, according to Alexander (2019), Cape Town introduced Day Zero: the scarcity of resources is not only an aggravating factor in the deterioration of the environmental situation, is inextricably linked to the emergence of conflicts both on a regional and global scale. The water reserves of the South African metropolis are nearly depleted due to planning errors and climate change. Population growth and severe drought intensified by climate change are, according to Welch (2018), at the root of one of the most dramatic urban water crises in the world. Helen Zille, former Cape Town mayor and the current premier of South Africa's Western Cape province wonders: when Day Zero arrives, how do we make water accessible and prevent anarchy?

In the past, according to Welch (2018) a dramatic situation of such magnitude in a cosmopolitan city was almost unthinkable. But overdevelopment, population growth, and climate change have altered the balance between demand and water availability, with the result that urban centres located from North to South America and from Australia to Asia are increasingly threatened by severe shortage of drinking water. In India, several big cities do not have enough water for everyone, and the water service operators in Melbourne, Australia, reported last summer that they could run out of water within a decade.

That brings even closer to the dystopian scenario envisaged by the participants of this study; paying attention to natural resources becomes a nonnegotiable imperative for homo resaliens. The rediscovered sense of limit is the key to achieving an environmental and political equilibrium.

The Hope for Another Political Space

According to Hawken et al. (2007), when a manufacturer understands that a supplier of basic components is overloaded with orders and will only deliver with delay, he immediately takes steps to ensure that his production cycle is not interrupted. Living systems, continues the scholars, which are basic components for the life of the planet, are overburdened and run the risk of no longer meeting orders. Until recently, the production world could ignore all this, because it did not affect its production process or its costs. What is consumed in the environment is not matter or energy but order and quality, that is, according to Hawken et al. (2007), structure, concentration, and purity of the material. To create economic value the production economy e extracts "quality" from nature.

It is therefore more interesting to worry about the specific qualitative aspects of natural capital, rather than questioning how long the stocks of physical resources will last. If relevant industries take structured matter away from the system concentrated and faster than the system uses to re-qualify it, and at the same time destroy the means for its reconstruction (ecosystems and habitats), a fundamental problem is introduced into production. Therefore, it is the decline of living systems that is the limiting factor that economic development must face today, in recognition of the reciprocal complementarity, and not substitutability, between artificial capital built by man and natural capital.

The inversion of the trend described above is nothing else than favouring investments in natural capital: a change of course with respect to today's reality, dominated by the prevalence of incentives for "disinvestment" in natural capital, rather than the opposite. It is therefore necessary to apply the ninth R proposed by Latouche (2007) which closes the political programme of degrowth: to decree a moratorium on technical and scientific innovation: redirect some areas of scientific research and respect natural and human limits. Restructuring the taxation systems with the aim of increasing the load on all those activities that damage the social and natural environment is a process that requires sufficiently long periods of consolidation, however, despite all difficulties, it is at the same time a path that allows to define a gradual and well-defined road for changes in innovation.

We must abolish linear processes and transform them into cyclical processes, as nature teaches us, so as not to waste anything because the waste of an organism is the nourishment of another. It seems interesting to propose a political programme which is a synthesis of desires, drives, vision, and literature analysed. Starting from the premise that technological innovation and the imperative of responsibility should delegate its values in order to make concrete the possibility of an economic system that takes into account living systems, we feel the need to integrate the 8Rs, Latouche's manifesto of Degrowth (2006) with other 8 reimagined Rs:

1. Remember that the environment is not a minor productive factor, it is the envelope that contains, supplies, and sustains our current economic system.
2. Rediscovering the functioning of natural capital. It represents life and therefore is outside the mercantilist logic as there is no possibility that it will be replaced.
3. Re-adjustment income inequalities.
4. Reinvest in human capital: it will increase the production of relational goods, such as cooperation, knowledge, mutual aid and so on: goods that do not know crises but those that can only increase.
5. Restore the political system: it is time to create a responsible governance agenda based on the needs of people and not on the needs of business, finance, and lobbyists.
6. Strengthen investment in qualitative research.
7. Reduce complexity.
8. Restructure the taxation systems.

The challenge that we have to embrace is a mental shift: the road towards *DemocraCity* is not a simple set of laws ready for use; it is a conceptual revolution, a change of our lifestyles. Only when this transformation has happened will it be possible to concretely elaborate the programmatic phase and draw up the political agenda of actions to be undertaken to achieve the concrete utopia.

In conclusion, the case studies of Transition Darebin and Fiorivano le viole, framed as a virtuous examples of transition initiatives, allow for the merging of theories such as degrowth, the application of resilience for community, cohousing, gift theory, sense of community. However, although the premises and the active participation of the members of such remarkable communities were passionate, although values such as trade (especially for TD) and arts (core value of FLV)—following the paradigm of McMillan—are vibrant within the communities under examination, there are limits in vision, lack of trust in institutions and in academia and a resistance to change, typical of a weak community. In order to answer one of the main questions that I have proposed during my interviews "Is it possible to consider Transition Darebin a community builder and a social thermometer in order to achieve the Buen Vivir?". Unfortunately, the answer is not entirely clear or positive.

References

Alexander, C. (2019). *Day zero, 365 days later*. https://thebulletin.org/

Becker, C. (2006). The human actor in ecological economics: Philosophical approach and research perspectives. *Ecological Economics., 60*, 17–23.

Benyus, J. M. (1997). *Biomimicry. Innovations inspired by nature*. William Morrow, New Nork.

Bertorello, M., & Corradi, D. (2011). *Capitalismo tossico. Crisi della competizione e modelli alternativi*. Edizioni Alegre.

Chatterton, P., & Cutler, A. (2008). *The rocky road to a real transition: The transition towns movement and what it means for social change*. The Trapese Collective.

Christian, D. L. (2003). *Creating a life together: Practical tools to grow ecovillages and intentional communities* (Kindle Ed.).

Commoner, B. (1974). *The closing circle; nature, man, & technology*. Bantam Books, New York.

Costanza, R., Cumberland, J., Daly, H., Goodland, R., & Norgaard, R. (1997). *"An introduction to ecological economics", International Society for Ecological Economics*. Lucie Press, Raton.

Crouch, C. (2000). *Coping with post democracy*. Fabian Pamphlets
Fromm, E. (1977). *To have or to be?*
Gare, A. (2000). Creating an ecological socialist future. *Capitalism Nature Socialism, 11*(2), 23–40.
Hawken, P., Lovins, A., & Lovins, L. H. (2001). *Capitalismo naturale*. Edizioni Ambiente.
Hawken, P., Lovins, A., & Lovins, L. H. (2007). *Capitalismo naturale. La prossima rivoluzione industriale*. Edizioni Ambiente.
Hopkins, R. (2008). *The transition handbook: From oil dependency to local resilience*. Chelsea Green Publishing.
Imanichi, T. (2009). *An introduction to eco-ethica*. UPA.
Kapp, K. W. (1991). *Economia e ambiente*. Saggi scelti, Otium, Ancona.
Kempf, H. (2010). *Per salvare il pianeta dobbiamo farla finita con il capitalismo*. Garzanti.
Kimhi, S., & Shamai, M. (2004). Community resilience and the impact of stress: Adult response to Israel's withdrawal from Lebanon. *IEEE Journal of Community Psychology, 32*(4), 439–451.
Latouche, S. (2007). *La scommessa della decrescita*. Feltrinelli.
Latouche, S. (2009). *Farewell to growth*. Polity Press.
Latouche, S. (2012). Can the Left Escape Economism? *Capitalism Nature Socialism, 23*(1), 74–78.
Latouche, S. (2013). Prefazione in Bonaiuti, M *La grande transizione: Dal declino alla società della decrescita*. Bollati Boringhieri, Torino.
Lietaert, M. (2007). *Cohousing e condomini solidali. Guida pratica alle nuove forme di vicinato e vita in comune con allegato documentario "Vivere in cohousing"*. Editrice Aam Terra Nuova.
Matthey, A. (2010). Imitation with intention and memory: An experiment. *Journal of Behavioral and Experimental Economics* (formerly *The Journal of Socio-Economics*), *39*(5), 585–594.
Maturana, H., & Varela, F. (1987). *L'albero della conoscenza*. Garzanti.
McDonough, W., & Braungart, M. (2002). *Cradle to cradle: Remaking the way we make things*. North Point Press, New York.
Morin, E. (1985). *Sociologia della sociologia*. Edizioni Lavoro, Roma.
Morin, E. (1999). *Seven complex lessons in education for the future*. UNESCO Publishing.
Munari, B. (1997). *Arte come mestiere*. Editore Laterza, Roma.
Muraca, B., & Döring, R. (2018). *From (strong) sustainability to degrowth: A philosophical and historical reconstruction*. Routledge.
Nakamura, T., & Kijima, K. (2009). System of system failures: Meta methodology for IT engineering safety. *Systems Research and Behavioral Science, 26*.

Norris, F. H., & Stevens, S. P., Pfefferbaum, B., & Wyche, K. F., & Pfefferbaum, R. L. (2008). Community resilience as a metaphor, theory, set of capacities, and strategy for disaster readiness. *IEE American Journal of Community Psychology, 41,* 127–150.

Pallante, M. (2007). *La decrescita felice. La qualità della vita non dipende dal pil.* Editori Riuniti.

Pauli, G. (2014). *Blue economy: 10 Anni, 100 Innovazioni, 100 Milioni di posti di lavoro.* Edizioni Ambiente.

Siebenhüner, B. (2000). Homo sustinens—Towards a new conception of humans for the science of sustainability. *Ecological Economics, 32*, 15–25.

Steffen, A. (2009). Planetary boundaries and the failure of environmentalism. http://www.worldchanging.com/archives/010551.html

von Glasersfeld, E., Stanzione, M., & Tagliagambe, S. (1994). *Quattro dialoghi su scienza e filosofia.* Odredek, Roma.

Welch, C. (2018). Why cape town is running out of water, and who's next. https://www.nationalgeographic.com/

CHAPTER 7

The Best of All Possible Worlds

A "Desirable" Way Out

This final chapter wants to take stock of this long multidisciplinary journey that sees a reality such as Transition Darebin (and FLV) as an example, a source of inspiration in which feelings and drives such as hope, respect, fear, inner conflicts (and perhaps failures) are intertwined making the road towards *DemocraCity* difficult, but it is still possible to reach this destination. In this study we tried to demonstrate that degrowth, the road to *DemocraCity*, does not spread and does not follow bottom-up dynamics, as suggested by Barbara Muraca and Döring (2018) and other scholars of degrowth.

The voluntary and conscious choice of ordinary people to change their daily life, the pursuit of frugality, the application of resilience for community (a term so abused in the academy environment that it has lost all meaning and effectiveness) demonstrate the applicability of the paradigm and the guidelines set out by Hopkins, rather than those outlined by Latouche and his 8Rs. Thus, this journey has refuted the conclusions of many scholars of the degrowth paradigm. Only through a careful political analysis supported by an academia finally disconnected from quantitative methods and from the industrial production of knowledge, will it become possible to talk about an intelligent technological innovation to reach, ten years after Latouche's degrowth paradigm, concrete utopia.

M. Carcea, *The Road Towards DemocraCity*,
https://doi.org/10.1007/978-981-16-3505-2_7

According to the authors presented on this journey, leaving the growth society behind is possible. Finding a new direction, or according to the members of Transition Darebin and Fiorivano le viole, a new narrative is the only key in order to progress and improve quality of life, guaranteeing respect for the Earth and for future generations. This is not only a good opportunity, but a moral imperative. The degrowth paradigm is the result of an evolutionary process of scientific maturation. In order to start what will lead us towards *DemocraCity*, we need a mental revolution that allows each individual to recognize the limits of the capitalist paradigm. The ideology of growth has created more limits and barriers despite its aim to break every frontier and allow unlimited growth, able to extend the power of man in everywhere and at all times, making him the absolute master of the cosmos.

The road to *DemocraCity* is a challenge: it is a cultural, social, economic and political revolution. It is a new way of thinking about nature, social cohesion, and one in which the quality of work and of democracy is needed. Taking care of people, building a community, as we saw in this case study, requires time. Even though this is considered wasted time from the point of view of productivity, this "freed time" is needed to achieve a society based on degrowth.

Slowness is a pivotal point. It is a positive feature which can be found in small realities of growth objectors and resilient communities such as Transition Darebin. A society based on degrowth is the concretization of the victory of the qualitative against the quantitative. Small communities, perhaps permeated by a vein of distrust that give an almost sectarian identity, certainly justified by the delicate balance on which they were built, have become "ready and resilient" in the face of globalized consumerism, the system of this disposable society.

The cohesion of individual growth objectors involves the establishment of small solidarity communities, capable of combining their strengths, their knowledge, their peculiarities, and diversity for the common purpose of becoming autonomous from the system of growth. They are groups that have well understood the sense of limit and the importance of the local dimension and who are able to climb back on the overturned boat. Although unaware of the paradigm of happy degrowth, of Serge Latouche assertions that the road to the redesign of the collective imaginary is still long and the bond of this paradigm with the academia is still profound, TD is a virtuous example of horizontal transition.

The sense of community, the self-production of basic necessities following the principle of Rob Hopkins and of the permaculture is vibrant. The regrowth of human and ecological values, arts, the trade, reuse, and recycling of products are examples of that enhanced resistance that we can find in Fix It or within the Repair Café, in order to fight planned obsolescence but it also implies a vision for a new narrative, fundamental feature of the homo resaliens.

Choosing to be ready and resilient means making oneself available to others, sharing one's know-how with others, not in exchange for money but of equal help in different areas. It is virtuous circle of services based on the gift paradigm and on reciprocity that enriches the life of members, allowing them to be ready to increase their know-how through initiative such as Transition in the Pub, learning new skills and forming bonds of friendship and mutual aid. Belonging to a transition initiative means strengthening the bonds of good neighbourliness: with activities taking place in common spaces members know each other, interact and help each other and in addition there is a positive impact on expenses, reducing individual consumption and costs in terms of overall pollution.

Backyards festivals, urban gardens, fruit swaps, and fruit banks are able to transform a sprawling metropolis such as Melbourne, expression of *turbocapitalism*, into a small, slow bucolic, and frugal reality in which individuals can grow and exchange seasonal products; a way to take care of the earth and regain space. Transformation, urban and cultural redevelopment is the challenge taken up by Fiorivano le viole. Again, *trade vs arts, The Transition Handbook vs Le Petit traité*. Alternatives such as Transition Darebin are now numerous all over the world and welcome more and more members: but not very often are they supported by political forces in the implementation of projects related to the environment and to the care of the territory.

A transition initiative should be a place which involves democracy and cooperation, giving rise to meetings and assemblies in which all members work towards the same goal: the change of the collective imaginary and living their own lives, differentiating between well-being and well-having. In addition, regaining creative control of one's live will reduce the neurotic frustration that characterizes capitalism.

Reconnection with the earth, one of the basic characteristics of initiatives such as Transition Darebin, is fundamental for the pillars that builds *DemocraCity*. Regrowth mentioned by the members of TD also begins with care for the territory and the rediscovery (concept shared with the members behind FLV) of those values that bind people to nature. Communities such as Transition Darebin are showing us that the concrete utopia, i.e. the creation of a desirable word through virtues such as cooperation and solidarity is achievable through positive and transversal relationships. However, are we sure that these examples are enough? In 2004 the Manifesto of Degrowth was the self-production of yogurt. Are sure that we are moving forward? Though admirable, all this is not enough.

In 2018, more than 10 years after the semantic bomb dropped by Latouche, academics, growth objectors, and degrowth enthusiasts are still convinced that degrowth can have indicators (and we were also convinced of this at the beginning of our journey), or that such paradigm can follow bottom-up dynamics. In particular, Barbara Muraca and Döring cites in her book *From (Strong) Sustainability to Degrowth: A Philosophical and Historical Reconstruction* (2018) the result of a survey conducted by Dennis Eversberg and Matthias Schmelzer within the paper: *The DegrowthSpectrum: Convergence and Divergence Within a Diverse and Conflictual Alliance. Environmental Values* (2018).

According to Muraca: "A recent questionnaire survey about motives, attitudes, and practices of grassroots activists within the degrowth spectrum has shown two main points of consensus: 1) the profound conviction that sustainability and growth are incompatible, and that growth is soon coming to an end in industrialized societies of the Global North, and 2) the idea that a positive social transformation "is critical of capitalism, pro-feminist, peaceful and bottom-up" (p. 34).

In addition, Muraca continues "following recent scholarly research, we maintain that the degrowth movement does not only call for a repoliticization and reconceptualization of sustainability, but also envisions a radical social-ecological transformation that is driven by the democratization of societal relations to nature and characterized by a bottom-up approach of a multitude of diversified and networked social experiments worldwide" (p. 36). Are we sure that it is truly like that? If so, this research path is not a viable one.

It was unable to find indicators, tools, and methods for the dissemination of such a mental boost within the local political representatives. The transition towards a degrowth society, at least in the examples of our case studies, are certainly characterized by optimism and principles of responsibility towards nature and future generations, however, it rather follows horizontal dynamics and characteristics of *weak communities*.

The hope of the growth objectors is that micro behaviours (this is the case thus of these initiatives) are new features introduced to escape from customary behaviour. However, if novelty is too expensive, does not have sufficient appeal, is not shared, or is opposed (as in the case of the political-economic theories under examination, opposed by both quantitative academia and political class), then no combinatorial effect commences, i.e. there is no contagion effect.

The informational only contagion is not sufficient to transform the experiment into custom, or to achieve macro behaviour result, and even if, at the beginning, the information is evenly distributed, and the results are visible to all members of a particular community. However, if a group of individuals randomly chooses an alternative path (more comfortable and easier as the current one), this alters the equidistributional weighting and leads the system to choose that specific alternative.

The challenge is the reconstruction of a collective imaginary based on dematerialization, a rediscovered sense of limits, political consciousness and technological innovation overseen by a new social contract. Biomimicry, investments in natural capital, productivity of resources are nothing if we do not reverse the "diabolical" pyramid, which has completely destroyed human and excluded natural capital.

Continuing to frame degrowth as a movement from below appears to be a quasi-de-responsibility of a quantitative academia which does not want to admit that the problem lies in a forgotten *paideia*. In order to achieve that, should we hope for a speedy rapprochement between the combatants which will improve the conditions of the exploited and the oppressed?

Between Present and Future

Remaining anchored to the liberal model contributes to the advance of the mentioned and feared Mad Max scenario. Crouch (2000) emphasized that the elections continue to take place and to condition governments, the electoral debate is a tightly controlled show, conducted by rival groups of experienced professionals using techniques of persuasion and is exercised using a limited number of issues selected by these groups. The mass of citizens, according to the data, plays a passive, acquiescent, even apathetic role, limiting itself to reacting to the signals it receives. Apart from the spectacle of the electoral struggle, politics is decided in private by the interaction between elected governments and elites that almost exclusively represent economic interests.

In these conditions, under the control of economic lobbies, policies advocating equality and the redistribution of wealth find little buy-in. This leads to a transformation of the left and the centre, in which the voice of the people decreases. A democratic path rather than being a circular movement becomes a parabolic movement and right now we are in the descend of its path. Data unfortunately confirm this. Psychological resistance towards politics, fear of corruption and lack of responsibility by our representatives nullify the work of reality such as transition initiatives.

On the other hand, political parties remain fundamental to avoid the anti-egalitarian tendencies of a post-democratic society, but we cannot be satisfied to work for our purposes only through a party: we must also work on a party from the outside, supporting initiatives such as TD that should be able to put pressure on the political class with their achievements. Those parties not under any pressure will remain rooted in day after scenario made by corporate lobbies, and the interest groups who attempt to act without reference to the building of strong left-wing parties will be overridden by business lobbies.

We need to keep in relation to each other these apparently conflicting forms of action, movements, and parties. The political class wants to exclude citizens as much as possible from any direct involvement in politics, but at the same time does not want to be ignored. This is the paradox of our time. The solution that sustains the idea of *DemocraCity* is to find answers in order to encourage the highest level of minimum participation. The committed citizen and advocate of egalitarianism will consider the same paradox from a different point of view, seeing the dependence of the political elite on a limited mass participation as an opportunity to find a maximum of opportunities for commitment. Thus, as if we attempt to close the circle, the word opportunity returns, a word that is inextricably linked to the one of crisis.

Caplan (1961) has argued that a crisis is a transitional moment, a moment in which everything suffers a sudden change, a time from which the individual comes out transformed. A crisis is a sort of *Janus Bifrons*, in which the binominal *danger/opportunity* can create new solutions or bring us to oblivion. Morin reminds us that a crisis is an extraordinary event that reveals the virtual, the unconscious, the possible as opposed to reality, the visible and the conscious. According to Maier (1996), capitalism is an economic system, extremely complex, and it is attributable to those systems characterized by a high intensity of complexity that coerce other systems hierarchically inferior. As Golinelli (2005) states, each institution of any order and degree, qualifies itself in obedience to the basic drivers of "systemic complexity".

A system, dynamic, adaptive, complex, and non-linear is conditioned by a combinatorial action exerted by the individuals who compose it, by their groups or communities, as well as by the wider and more complex systems of which it is a part. A crisis, thus, becomes a pillar of a society because it transforms the systemic complexity of the whole society. Perrow (1999) argues that systemic complexity makes malfunctions and failures inevitable. It may be argued that a crisis is an endogenous element of our current dominant imaginary. It is outright the thought of Ferrer-Pacces (1970, p. 41): "the system has created and nurtured for centuries the seeds of its self-destruction".

The term obsolescence, extremely dear to degrowth and transition scholars, should not be applied only to goods and markets; homo resaliens must develop the ability to understand that a crisis is the result of endogenous factors such as institutional fragility, political and regulatory obsolescence, lack of transparency and adequacy of information. This nexus leads us to another contradiction of the dominant logic based on *laissez faire* ideology. Capitalism was conceived and is still understood by technologists, as a rational process, a self-expanding and self-balancing system, in which crisis is completely ignored, is irrelevant or otherwise can be overcome through optimism and undisguised faith in the technological progress.

Between Emergency and Urgency

The famous view of the gulf of Naples profiled by the majestic silhouette of Vesuvius has now been replaced in the world by that of a city reduced to a huge open-air dump. Naples is periodically submerged by waste, the cycle of which is managed by organized crime with the acquiescence of a corrupt political class. By evading the disposal and control procedures, toxic and poisonous materials arrive from the north as undifferentiated common garbage which, buried in the soil, poison the aquifers, agricultural products, livestock, and their derivatives, starting with those products that always constitute the excellence and identity of our food tradition.

In August 2010 in China, a 100 km-long queue of vehicles formed on the highway leading to Beijing: it took a month to dispose of it with the intervention of four hundred police to keep order. In Tehran in December of the same year, schools, universities, public offices, and banks were ordered to close due to the very high rate of pollution in the air and the use of masks was imposed on those who were obliged to carry out activities open.

Despite the numerous and alarming reports of climate change mutations and overwhelming evidence produced by a growing breathlessness of the world, from desertification to the erosion of biodiversity, from growing scarcity of water to land degradation, from acidification of the oceans to extension of urban centres, from temperature rise to melting glaciers, rulers and international institutions continue to confirm the diagnosis i.e. the awareness of the ecological catastrophe is too slow to avoid the worst. Indeed, as we argued, unfortunately, the problems resulting in a catastrophe can no longer be avoided, therefore, the only option is to ask how to manage it.

Even if, from one day to the next we ceased the production of the emissions of greenhouse gases, pollutants, the unbridled withdrawal of resources of all kinds and their senseless waste, if we reduced, in short, our ecological footprint to a sustainable level, we would still have a temperature rise of two degrees by the end of the century. This effect will result in increasingly serious food problems and in the scarcity of drinking water for many populations, in the desertification of ever larger areas of the planet, in coastal areas submerged by water from melting glaciers with hundreds of millions of environmental refugees.

Moreover, without the reckless destruction of mangroves, rampant real estate speculation, and international tourism, the 2004 tsunami in Southeast Asia would not have caused so much destruction and more than two hundred thousand deaths. Climate change and the reduction of forests with the alteration of the habitats of many animal species put wild animals in contact with humans more and more easily, and even closer contact when these animals are captured to be sold in crowded markets, making easier to jump species for their pathogens.

Furthermore, the farms, in particular of chickens and pigs, with concentrations of many animals in small spaces, fed with feed containing antibiotics, favour a strong selective pressure on their viruses and bacteria, which rapidly mutate towards more aggressive strains and types also towards the species human, as was the case for avian and swine flu. A further contribution, as Quammen reminded us in 2012, to the spread of pathogens, is then given by globalization, which, due to the frenetic transfer of people and goods to every part of the planet, favours the transition from epidemics to pandemics.

As we already discussed, in 1972, the Club of Rome with its *The Limits to Growth* stated that if the world population continued to grow at the rate of those years, the growing demand for food would have impoverished the fertility of the soil, the growing production of goods would have increased the pollution of the environment, the depletion of the reserves of natural resources would have provoked conflicts for their conquest; diseases, epidemics, hunger, conflicts would have held back population growth.

This pandemic can be a useful warning, to avoid new and more serious pandemics. Covid-19 is a reaction to the state of stress we have caused to the planet and therefore to prevent new similar events we must reduce environmental alterations, such as loss of biodiversity, alteration of habitats and climate change, favouring industrial and agricultural production processes based on the circular economy, sustainable, with recourse to renewable energy sources.

However, the pandemic has also highlighted shortcomings in national health systems, especially in those countries where it has been decided to dismantle the public system: reversing this trend and adequately financing public health structures, together with prevention policies, will be a fundamental barrier to future pandemics.

Conclusion: The Detoxification of the Imagination

We can argue that complexity represents a way of overcoming certain difficulties, however, on the other hand, a form of feeding further disasters, precisely through integration, coordination, communication between systemic components. Complication and complexification are the (infamous) tools of our current (political) economic system. From J.M. Keynes and other fathers of regulated *capitalism*—such as Karl Polanyi and Joseph Schumpeter—complication is based on the increase of variables and constraints under which the system and its agents must operate. Complexification instead implies the role of technology and the rules related to it; such an answer to crisis triggers phenomena and self-controlled behaviours between the agents and makes the system more flexible (an emblematic example was the aforementioned Solow and his response to Limits to Growth).

Both the instruments so far implemented, in order to overcome the impasse, proved to be fallacious (inefficient and ineffective) and valid perhaps in the medium term and tending, as history teaches us (the famous "30 glorious years" recalled by Engels), to replication. Degrowth, a reimagined degrowth or regrowth, through its intervention in the culture inherent in the system, will be able to instil different behaviours and convictions which could heal the crisis and avoid it in the future. It is a slow path, but the goal is reducing complexity, which has a different meaning to simplification.

Reducing complexity is a mental shift, a countermeasure probably slower (compared to the complication and complexification), and definitively more expensive, however, it completely changes the hierarchy of a system (wealth distribution and access to natural heritage between the Global North and Global South and, within each society, between classes, individuals and future generations) and the perspectives of all agents of such a system. In the previous chapter we tried to reimagine the Rs of Latouche, in particular our fifth stressed the renovation of our political system in which a responsible governance agenda based on the needs of people, and not on the needs of the business, finance, and lobbies took centre stage.

In light of data, drives, and aspirations that we found within the field of research we would like to propose these questions:

Would it be possible to apply a small, fixed sum to the income of each citizen and assign it to the political party chosen by the citizen himself? Could the same thing be applied to groups and associations such as TD or FLV? In addition, would it be possible to establish an assembly of citizens that is in charge for a year and formed by the most involved citizens who see the bills and can approve or reject them? (Also putting in systems to avoid lobbying or external pressure in general on the components of the assembly). These two proposals want to embrace the aforementioned reduction of complexity approach, such ideas want to drive ordinary people, such as the member of transition initiatives to action and within the political choice: a potentially large number of citizens could participate over time in these assemblies and bring with them a legacy of political commitment or at least a knowledge of that sense of limit experienced within the local communities.

Democrats should seek to strengthen the role of local and regional authorities, to advance the cause of decentralization and to protect and broaden the scope of citizenship services for which the local administration is responsible. The pandemic has also highlighted shortcomings in national health systems, especially in those countries where it has been chosen to dismantle the public system: reversing this trend and adequately financing public health structures, together with prevention policies, will be a fundamental barrier to future pandemics.

Reducing the complexity brings us to the example of Illich's snail, the original symbol of degrowth: accepting the natural limits of the biosphere, enjoying the slowness of life, rediscovering the beauty and taste of the world, freeing oneself from anxiety in order to find the reflective and contemplative moment that allows ourselves to increase well-being, but above all, rediscovering nature as an inventor in order to pursue a newfound agape in which a triangulation of ethos, a regrowth of ethical principles, and a virtuous imitation must be the basis for a true application of resilience and a meaningful concept of sustainability.

References

Caplan, G. (1961). *Prevention of mental disorders in children: Initial explorations*. Tavistock.

Crouch, C. (2000). *Coping with post democracy*. Fabian Pamphlets

Ferrer-Pacces, F. M. (1970). MNK. Manifesto del neocapitalismo. *L'impresa, gennaio-febbraio*.

Golinelli, G. M. (2005). *L'approccio sistemico-vitale al governo dell'impresa, vol. I, L'impresa* sistema vitale. Cedam.

Maier, M. (1996). Architecting principles for systems of systems. In *Proceeding of the Sixth Annual International Symposium. International Council on Systems Engineering*. Boston.

Muraca, B., & Döring, R. (2018). *From (strong) sustainability to degrowth: A philosophical and historical reconstruction*. Routledge.

Perrow, C. (1999). *Normal accidents living with high-risk technologies*. Paperbacks.

Glossary, Definitions, and Key Words

Degrowth: a cultural process of political and social transformation favourable to the controlled, selective, and voluntary reduction of economic production and consumption.

Resilience: word that derives from the Latin "resalio", iterative of "salio" (go up, rise again).

Homo resaliens: the one who has understood the sense of one's limits, the one who is able to arise from an overturned boat (metaphor of our capitalist society).

Transition Town: movement designed to prepare communities to deal with scenarios such as global warming and peak oil. The Transition Town movement was founded in Kinsale, Ireland, and in Totnes, England by environmentalist Rob Hopkins in 2005 and 2006, respectively.

***DemocraCity*:** utopian city, based on the principles of degrowth that takes its inspiration from the ideas of Campanella and later Kelsen

Blue economy: a global economy model dedicated to creating a sustainable ecosystem by transforming previously wasted substances into profitable goods.

Diabolic carousel: to allow the consumer society to continue its diabolic carousel, three ingredients are needed: advertising, which creates the desire to consume, credit, which provides the means, and accelerated and programmed obsolescence of products, which renews its need.

M. Carcea, *The Road Towards DemocraCity*,
https://doi.org/10.1007/978-981-16-3505-2

Keywords: degrowth, egrowth, resilience, alternative currencies, Transition Town movements, E.E. (Environmental Economics), Bioeconomics, inclusive democracy, biomimicry, eco-socialism, ethics, political alliances, imaginaries, sustainable degrowth economic crisis, politics, sustainability, Homo economicus, Homo resaliens, Gift theory, history of Ecological Economics thought, Club of Rome, voluntary simplicity, well-being, well having, political economy, climate change, global warming, ecological man, utopia, technology and governance, conviviality, complexity, socio-economic system, collective identity, social hierarchy, economic resilience, participative democracy, development, emergence, networks, ecology, reciprocity, mutual aid, economic growth, communities.

Reference

"The economics of degrowth". *Ecological Economics*. https://degrowth.org/wp-content/uploads/2012/11/Kallis_2012_The-economics-of-degrowth.pdf

Adkisson R. V. (2009). The economy as an open system: An institutionalist framework for economic development. In T. Natarajan, W. Elsner, & S. Fullwiler (Eds.), *Institutional analysis and praxis*. New York, NY: Springer. https://doi.org/10.1007/978-0-387-88741-8_2

Alexander, C. (2019). *Day zero, 365 days later*. https://thebulletin.org/

Alexander, S. (2012). Peak oil and the twilight of growth. *A lternative Law Journal, 37*(2), 86–90.

Ali, M. (2013). *Climate change impacts on plant biomass growth*. Springer.

Althabe, G. (1969). *Oppression et libération dans l'imaginaire*. Les communautés villageoises de la côte orientale du Madagascar; Maspero.

Amartya, S. (1982). *Poverty and famines: An essay on entitlement and deprivation*. Oxford University Press, Oxford.

Andrew, B. (2008). Market failure, government failure and externalities in climate change mitigation: The case for a carbon tax. *Special Issue: Symposium on Climate Change, Governance and Environmental Services,28*(5), 393–401.

Andreoni, V., & Galmarini, S. (2014). How to increase well-being in a context of degrowth. *Futures, 55*, 78–89.

Anguelovski, I. (2015). Tactical developments for achieving just and sustainable neighborhoods: the role of community-based coalitions and bottom-to-bottom networks in street, technical, and funder activism. *Environment and Planning C: Government and Policy, 33*(4), 703–725. https://doi.org/10.1068/c12347

M. Carcea, *The Road Towards DemocraCity*,
https://doi.org/10.1007/978-981-16-3505-2

Arendt, H. (2004). *Responsabilità e giudizio, a cura di Jerome Kohn*. Biblioteca Einaudi, Torino.

Arrow, K., et al. (1995). Economic growth, carrying capacity, and the environment. *Ecological Economics, 15*(2), 91–95.

Arthur, W. B. (2009). *The nature of technology: What it is and how it evolves*. Free Press.

Australian Bureau of Statistics. (2018). Annual Report, 2018–19. https://www.abs.gov.au

Ayres, R., & Warr, B. (2010) *The economic growth engine: How energy and work drive material prosperity (The International Institute for Applied Systems Analysis)*. Edward Elgar.

Bachrach, K. M., & Zautra, A. J. (1985). Coping with a community stressor: The threat of a hazardous waste facility. *Journal of Health and Social Behavior,26*, 127–141.

Badiou, A. (2006). *Il secolo*. Feltrinelli.

Bardi, U. (2007). Energy prices and resource depletion: Lessons from the case of whaling in the nineteenth century. *Energy Sources*, Part B, *2*(3), 297–304.

Bardi, U., & Lavacchi, A. (2009). A simple interpretation of Hubbert's model of resource exploitation. *Energies, 2*(3), 646–661.

Bardi, U. (2016). What future for the anthropocene? A biophysical interpretation. *Biophysical Economics and Resource Quality,1*(1), 1–7.

Barkemeyer, R., Holt, D., Preuss, L., & Tsang, S. (2014). What happened to the 'development' in sustainable development? Business guidelines two decades after Brundtland. *Sustainable Development, 22*(1), 15–32.

Barret, G. (2012). Explorations in the economics of aging. *Economic Record, 88*(281), 290–292.

Bauman, Z. (2000). *Liquid modernity*. Cambridge.

Bauman, Z. (2007). *Consuming life*. Cambridge.

Bauman, Z. (2009). *Living on borrowed time: Conversations with citlali rovirosa-madrazo*. Cambridge.

Bauman, Z. (2010). *Capitalismo parassitario*. Il Nocciolo.

Becker, C. (2006). The human actor in ecological economics: Philosophical approach and research perspectives. *Ecological Economics,60*, 17–23.

Beckerman, W. (1974). *In defence of economic growth*. J. Cape.

Beckert, J. (2007). *The great transformation of embeddedness: Karl Polanyi and the new economic sociology* (MPIfG Discussion Paper 07/1).

Belpomme, D. (2004). *Ces maladies crées par l'homme*. Albin Michel.

Benyus, J. M. (1997). *Biomimicry. Innovations inspired by nature*. William Morrow, New Nork.

Berg, B. L. (2004). *Qualitative research methods for the social sciences* (Vol. 5). Pearson.

Berger, S. (2008). K. William Kapp's theory of social costs and environmental policy: Towards political ecological economics. *Ecological Economics,67*(2), 244–252.

Berthoud, A. (2005). *Una philosophie de la consommation. Agent économique et sujet moral*. Presses univerisitaires du Septention, Villeneuve d'Ascq.

Bertorello, M., & Corradi, D. (2011). *Capitalismo tossico. Crisi della competizione e modelli alternativi*. Edizioni Alegre.

Besset, J.-P. (2006). *La scelta difficile. Come salvarsi dal progresso senza essere reazionari*. Dedalo.

Besset, J.-P. (2007). *La scelta difficile. Come salvarsi dal progresso senza essere reazionari*. Dedalo.

Benyus, J. M. (1997). *Biomimicry: Innovations inspired by nature*. William Morrow.

Bevilacqua, P. (2011). *Il grande saccheggio*. Laterza.

Bianchi, J. (2011). Overborrowing and systemic externalities in the business cycle. *American Economic Review,1*(7), 3400–3426.

Bloch, E. (1964). *The spirit of Utopia* (A. A. Nassar, Trans. [2000]). Standford University Press.

Billig, M. (2012). Undisciplined beginnings, academic success, and discursive psychology. *British Journal of Social Psychology,51,* 413–424.

Bloemmen, M., Bobulescu, R., Le Tuyen, N., & Vitari, C. (2015). Microeconomic degrowth: The case of community supported agriculture. *Ecological Economics,112*, 110–115.

Bogdanov, A. (1921). *Očerki Vseobščej Organizacionnoj Nauki* [Essays in the Universal Organizational Science]. Samara.

Bogdanov, A. (1980). *Essays in tektology: The general science of organization* (G. Gorelik, Trans. [1980]). Intersystems Publications, Seaside.

Bogdanov, A. (1984). *Red star—The first Bolshevik Utopia*. Indiana University Press.

Bornestein, J. (2013). What to save? The normative dilemmas of resilience. *Centre for Human and Nature*, *6*(3), 7.

Boulding, K. (1966). The economics of knowledge and the knowledge of economics. *American Economic Review, 56*(1/2), 1–13.

Boulding, K. (1992). *Towards a new economics: Critical essays on ecology, distribution, and others theme*. Edward Elgar.

Boulding, K. (1993). *The structure of a modern economy: The United States, 1929–89*. Macmillan.

Breton, M. (2001). Neighborhood resiliency. *IEEE Journal of Community Practice, 19*(1), 48–61.

Brougher, K. (2015). Art and nuclear culture. *Bulletin of the Atomic Scientists,69*(6), 11–18.

Burke, P. J., Shahiduzzaman, M., & Stern, D. I. (2015). Carbon dioxide emissions in the short run: The rate and sources of economic growth matter. *Global Environmental Change,33*, 109–121.

Cacciari, P. (2006). *Pensare la decrescita. Sostenibilità ed equità*. Intra Moenia.

Caillé, A . (1998). *Il terzo paradigma. Antropologia filosofica del dono*. Bollati Boringhieri.

Caillé, A. (2000). *Anthropologie du don*. Desclée de Brouwer.

Caillé, A. (2005). Esquisse d'une analytique du don. *Dé-penser l'économique*. La Découverte/MAUSS.

Caillé, A. (Ed.). (2007). *La quête de reconnaissance: Nouveau phénomène social total*. La Découverte/MAUSS.

Caillé, A. (2009). *Théorie anti-utilitariste de l'action*. La Découverte/MAUSS.

Caillé, A., & Chanial, P. (2008). L'amour des autres: *Care*, compassion, et humanitarisme. *Revue du Mauss,2*(32), 97–124.

Calafati, A. G. (1997). Il degrado ambientale come fallimento del mercato? *La questione agraria*, 67.

Caplan, G. (1961). *Prevention of mental disorders in children: Initial explorations*. Tavistock.

Caradonna, J. (2018). *Routledge handbook of the history of sustainability* (pp. 339–362). Routledge.

Cardoso Machado, N. M. (2011). Karl Polanyi and the new economic sociology: Notes on the concept of (Dis)embeddedness. *RCCS Annual Review, 90*(3), 19–140.

Carson, R. (1962). *Silent spring*. Houghton Mifflin.

Castoriadis, C. (1975). *L'Institution imaginaire de la société*. Seuil.

Castoriadis, C. (1996). *La montée de l'insignifiance* (p. 96). Les carrefours du labyrinthe IV.

Castoriadis, C. (1998). *L'enigma del soggetto*. Dedalo.

Castoriadis, C. (2005). *Une sociétè à la dérive* (p. 244). Seuil.

Castoriadis, C. (2010). *Relativismo e democrazia. Dibattito con il MAUSS*. Elèuthera.

Chanial, P. (2008). *La société vue du don: Manuel de sociologie anti-utilitariste appliquée*. La Découverte/MAUSS.

Chapman, I. (2014). The end of peak oil? Why this topic is still relevant despite recent denials. *Energy Policy,64*, 93–101.

Chatterton, P., & Cutler, A. (2008). *The rocky road to a real transition: The transition towns movement and what it means for social change*. The Trapese Collective.

Christian, D. L. (2003). *Creating a life together: Practical tools to grow ecovillages and intentional communities* (Kindle Ed.).

Christian, D. L. (2007). *Finding community: How to join an ecovillage or intentional community* (Kindle Ed.).

Chochet, Y. (2009). *Antimanuel d'écologie*. Breàl – Parution.
Club of Rome (2004). *The limits of growth: The 30-year update*. Chelsea Green.
Cochet, Y. (2005). *Apocalisse dell'olio*. Parigi, Fayard.
Coluccia, P. (2001). *La banca del tempo*. Bollati Boringhieri.
Comeliau, C. (2003). *Brouillons pour l'avenir: contributions au débats sur les alternatives, Iued*. Puf.
Commoner, B. (1974). *The closing circle; nature, man, & technology*. Bantam Books, New York.
Commoner, B. (1986). *Il cerchio da chiudere. Presentazione di Giorgio Nebbia Appendice di Virginio Bettini*. Garzanti Libri.
Conti, L. (1987). *Influenze e tendenze nello sviluppo della cultura ecologica*. Russo e Silvestrini.
Costanza, R. (1991). *The science and management of sustainability: Ecological economics*, Columbia University Press, New York.
Costanza, R., Cumberland, J., Daly, H., Goodland, R., & Norgaard, R. (1997). An introduction to Ecological Economics. *International Society for Ecological Economics*, Boca St. Lucie Press, Raton (Florida).
Crouch, C. (2000). *Coping with post democracy*. Fabian Pamphlets
Crouch, C. (2008). What Will Follow the Demise of Privatised Keynesianism? *The Political Quarterly, 79*(4), 476–487.
Dahlman, C. J. (1979). The problem of externality. *The Journal of Law and Economics,22*(1), 141–162.
D'Alisa, G., Demaria, F., & Kallis, G. (2014). *Degrowth: A vocabulary for a new era*. Routledge.
Dacrema, P. (2010). *Il miracolo dei soldi. Come nascono, dove vanno, come si moltiplicano*. Etas.
Daily, G., Postel, S., Bawa, K. S., & Kaufman, L. (1997). *Nature's services: Societal dependence on natural ecosystems*. https://www.researchgate.net/publication/37717461_Nature's_Services_Societal_Dependence_On_Natural_Ecosystems
Daly, E. H. (1977). *Steady-state economics*. Island Press, Washington.
Daly, H. (1981). *Lo stato stazionario. L'economia dell'equilibrio biofisico e della crescita morale, Sansoni*. Firenze.
Daly, H. E. (1992). Allocation, distribution, and scale: towards an economics that is efficient, just, and sustainable. *Ecological Economics, 6*(3), 185–193.
Daly, H. E. (2010). *Ecological economics: Principles and applications*. Island Press.
Davis, C. (2012). The politics of 'fracking': Regulating natural gas drilling practices in Colorado and Texas. *Review of Policy Research,29*(2), 177–191.
De Benoist, A. (2006). *Ultimo anno. Diario di fine secolo*. Edizioni Settecolori.
Deese, R. S. (2009). The artefact of nature: 'Spaceship Earth' and the dawn of global environmentalism. *Endeavour, 33*(2), 70–75.

Dempsey, J. & Robertson, M. (2012). *Ecosystem services: Tensions and developments within neoliberal environmentalism. Progress in Human Geography.*

Development Co operation Report. (2018). *Joining forces to leave no one behind.* https://www.oecd-ilibrary.org/

de Waal, F. (2006). *Le Singe en nous.* Fayard.

Drexhage, J., & Murphy, D. (2010). *Sustainable development: From Brundtland to Rio 2012.* United Nations Headquarters, 1st Meeting by the High-Level Panel on Global Sustainability, United Nations, New York.

Eco, U. (2004). *Storia Della Bellezza.* Rizzoli.

Eco, U., & Fedriga R. (2004). *Storia della Filosofia.* Edizioni Laterza.

Ehrlich, P., & Ehrlich, A. (1969). *The population bomb.* Ballantine.

Ehrlich, P., & Holdren, J. (1971). The impact of population growth. *Science, 171*, 1212–1217.

Esposito, M. T., & Soufani, F. (2018). The circular economy: An opportunity for renewal, growth, and stability. *Thunderbird International Review, 60*(5), 725–728.

Eversberg, D., & Schmelzer, M. (2018). The degrowth spectrum: Convergence and divergence within a diverse and conflictual alliance. *Environmental Values,27*(3), 245–267.

European Community. (1993). *Fifth action programme on the environment.* Bruxelles.

Faber, B., Petersen, T., & Schiller, J. (2002). Homo oeconomicus and homo politicus in ecological Economics. *Ecological Economics,40*, 323–333.

Ferrer-Pacces, F. M. (1970). MNK. Manifesto del neocapitalismo. *L'impresa, gennaio- febbraio.*

Fetzer, T 2014, *Fracking growth* (CEP Discussion Papers, CEPDP1278). Centre for Economic Performance, London School of Economics and Political Science, London, UK.

Fisher, W. F., & Ponniah, T. (2003). *Another world is possible: Popular alternatives to globalization at the world social forum.* Zed Books.

Fistetti, F. (2008). Is a society of recognition possible? Interview with N. G. Canclini, A. Honneth, and A. Caillé, *Revue du MAUSS, 32*, 411–432.

Fotopulos, T. (1999). *Per una Democrazia Globale.* Eleuthera, Milano.

Fourastie, J. (1980). *Les Trente glorieuses ou la Révolution invisible de 1946 à 1975. Le Livre De PocheThings.* New York: North Point Press.

Fromm, E. (1977). *To have or to be?*

Fukuoka, M. (2009). *The one-straw revolution: An introduction to natural farming.* Review Books Classics.

Gadrey, J., & Jany-Catrice, F. (2005). *Les nouveaux indicateurs de richesse.* La Decouverte, Reperes edition.

Gadrey, J., & Jany-Catrice, F. (2006). *The new indicators of well-being and development.* Palgrave.

Galtung, J. (1969). Violence, peace, and peace research. *Journal of Peace Research, 6*(3), 167–191.
Gardner, H. (2006).*Multiple intelligences: New horizons*. Basic books.
Gardner, H. (2011). *Frames of mind: The theory of multiple intelligences*. Basic Books.
Gare, A. (1995). *Postmodernism and the environmental crisis*. Routledge.
Gare, A. (2000). Aleksandr Bogdanov's history, sociology and philosophy of science. *Studies in History and Philosophy of Science, Part A, 31*(2), 231–248.
Gare, A. (2000). Creating an ecological socialist future. *Capitalism Nature Socialism, 11*(2), 23–40.
Gare, A. (2002). Towards an inclusive democracy: The crisis of the growth economy and the need for a new liberatory project, by Takis Fotopoulos. *Review of Radical Political Economics, 34*(1) (34), 97–99.
Gare, A. (2016). *The philosophical foundations of ecological civilization: A manifesto for the future*. Routledge.
Georgescu-Roegen, N. (1971). *The entropy law and the economic process*. Harvard University Press.
Giovannini, E. (2012). Indagine conoscitiva sull'individuazione di indicatori di misurazione del benessere ulteriori rispetto al PIL. *Audizione del Presidente dell'Istituto nazionale di statistica Enrico Giovannini, Commissione V "Bilancio, tesoro e programmazione" della Camera dei Deputati*.
Gist, R., & Lubin, B. (1989). *Psychosocial aspects of disasters*. Wiley.
Glasson, J. (2012). *Introduction to environmental impact assessment*. Routledge.
Gluckman, M. (2006). *Politics, law and ritual in tribal society*. Aldine Transaction.
Godbout, J. T. (1998). *Il linguaggio del dono*. Bollati Boringhieri.
Goldsmith, E. (1972). A blueprint for survival. *The Ecologist*, n. 2.
Goldthorpe, J. (1998). Rational action theory for sociology. *The British Journal of Sociology, 49*(2), 167–192.
Golinelli, G. M. (2005). *L'approccio sistemico-vitale al governo dell'impresa, vol. I, L'impresa sistema vitale*. Cedam.
Golub, R., & Townsend, J. (1977). Malthus, multinationals and the club of Rome. *Social Studies of Science, 7*(2), 201–222.
Gorz, A. (1992). *Capitalismo, socialismo, ecologia*. Manifestolibri.
Graeber, D. (2011). *Debt: The first 5,000 years*. Melville House.
Gruchy, A. G. (1947). *Modern economic thought*. Prentice Hall.
Gruzinski, S. (1988). *La colonisation de l'imaginaire. Sociétés indigènes et occidentalisation dans le Mexique espagnol XVIè-XVIIIè siècle*. Gallimard.
Hardin, G. (1968). The tragedy of the commons. *Science, 162,* 1243–1248
Hawken, P., Lovins, A., & Lovins, L. H. (2001). *Capitalismo naturale*. Edizioni Ambiente.

Hawken, P., Lovins, A., & Lovins, L. H. (2007). *Capitalismo naturale. La prossima rivoluzione industriale*. Edizioni Ambiente.

Heidenreich, R. (1998). Economics and institutions: The socioeconomic approach of K. William Kapp. *Journal of Economic Issues,32*(4), 965–984.

Heidenreich, M. (2016). *Exploring inequality in Europe. Diverging income and employment opportunities in the crisis*. Edward Elgar Publishing Limited.

Heikkurinen, P. (2018). Degrowth by means of technology? A treatise for an ethos of releasement. Part 2, *Journal of Cleaner Production*, *197*, 1654–1665, Elsevier.

Helmreich, S. (2011). From spaceship earth to Google Ocean: Planetary icons, indexes, and infrastructures. *Social Research*, *78*(4), 1211–1242.

Helmsing, B. (2001). Externalities, learning and governance: New perspectives on local economic development. *Development and Change,32*(2), 277–308.

Hemmingsen, E. (2010). At the base of Hubbert's Peak: Grounding the debate on petroleum scarcity. *Geoforum,41*(4), 531–540.

Hertzberg, M., Siddons, A., & Schreuder, H. (2017). Role of greenhouse gases in climate change. *Energy & Environment,28*(4), 530–539.

Hirsch, R. L., Bezdek, R., & Wendling, R. (2005, February). *Peaking of world oil production: Impacts, mitigation, & risk management*. Science Applications International Corporation, U.S. Department of Energy, National Energy Technology Laboratory.

Hopkins, R. (2008). *The transition handbook: From oil dependency to local resilience*. Chelsea Green Publishing.

Hopkins, R. (2009). *Manuale pratico della transizione: Dalla dipendenza dal petrolio alla forza delle comunità locali*. Arianna, Bologna.

Höhler, S. (2007). Carrying capacity—The moral economy of the coming spaceship earth. *Atenea: A Bilingual Journal of the Humanities and Social Sciences XXVI*, 1, 59–74.

Holmgren, D. (2002). *Permaculture: Principles and pathways beyond sustainability*, Holmgren design service. Hepburn.

Holmgren, D. (2010). *Permacultura. Dallo sfruttamento all'integrazione: Progettare modelli di vita etici, stabili e sostenibili*. Arianna.

Holgrem, D., & Mollison, B. (1990). *Permaculture one: A perennial agricultural system for human settlements*. Tagari Publications.

Hubbert, M. K. (1949). Energy from Fossil Fuels. *American Association for the Advancement of Science, 109*(2823), 103–109.

Hubbert, M. K. (1959). Techniques of prediction with application to the petroleum industry. *American Association of Petroleum Geologists, 43*(7), 1767–1768.

Illich, I. (1973). *Energy and equity*. Marion Boyars.

Illich, I. (1992). *La convivialità*. Red, Como.

Illich, I. (2000). *Deschooling society*. Marion Boyars.

Illich, I. (2001). *Tools for conviviality*. Marion Boyars.
Illich, I. (2005). *La convivialità, Una proposta libertaria per una politica dei limiti allo sviluppo*. Boroli.
Imanichi, T. (2009). *An introduction to eco-ethica*. UPA.
IMF. (2019). Annual Report 2019. https://www.imf.org
Inman, M. (2014). Natural gas: The fracking fallacy. *Nature,516*(7529), 28–30.
IPCC. (2014). Climate change 2014: Synthesis report. Contribution of working groups I, II and III to the fifth assessment report of the intergovernmental panel on climate change [Core writing team, R. K. Pachauri and L. A. Meyer (eds.)]. IPCC, Geneva, Switzerland, 151.
Jackson, T. (2009). *Prosperity without Growth. Economics for a Finite Planet*. Earthscan.
Jackson, T. (2011). *Prosperity without growth: Economics for a finite planet*. Earthscan, London.
Jackson, T. (2013). New economic model needed not relentless consumer demand. *The Guardian Blog*.
Jackson, T., & Webster, R. (2016). Limits revisited. *A review of the limits to growth debate*. Creative Commons.
Jaffe, A. B., Newell, R. G., & Stavins, R. N. (2005). A tale of two market failures: Technology and environmental policy. *Ecological Economics, 54*(2–3), 164–174, Elsevier.
Jonas, H. (1985). *The imperative of responsibility: In search of an ethics for the technological age*. University of Chicago Press.
Jones, P. (2010). Responding to the ecological crisis: Transformative pathways for social work education. *Journal of Social Work Education,46*(1), 67–84.
Josselson, R. (2004). The hermeneutics of faith and the hermeneutics of suspicion. *Narrative Inquiry,14*(1), 1–28.
Kallis, G. (2011). In defence of degrowth. *Ecological Economics,70*, 873–880.
Kallis, G. (2017). Economics without growth. In M. Castells (Ed.), *Another economy is possible: Culture and economy in a time of crisis* (pp. 34–54), Cambridge University Press.
Kallis, G., Kerschner, C., & Martinez-Alier, J. (2009). The economics of degrowth. *Ecological Economics,84*, 172–180.
Kallis, G. Martinez-Alier, J., & Schneider, F. (2010). Growth, recession or degrowth for sustainability and equity? *Journal of Cleaner Production*.
Kallis, G., Kerschner, C., & Martinez-Alier, J. (2012). The economics of degrowth. *Ecological Economics*. https://degrowth.org/wp-content/uploads/2012/11/Kallis_2012_The-economics-of-degrowth.pdf
Kallis, G., Kostakis, V., Lange, S., Muraca, B., Paulson, S., & Schmelzer, M. (2018). Research on degrowth. *Annual Review of Environment and Resources*.
Kapp, K. W. (1950). *The social costs of private enterprise*. Harvard University Press.

Kapp, K. W. (1971). L'economia come sistema aperto. *Economia e ambiente, Saggi scelti*. Otium Edizioni, Ancona.
Kapp, K. W. (1991). *Economia e ambiente*. Saggi scelti, Otium, Ancona.
Karl, T. R., & Trenberth, K. E. (2003). Modern global climate change. *Science,302*, 1719–1723.
Kempf, H. (2007). *Comment les riches dètruisent la planetè*. Le Seuil.
Kempf, H. (2010). *Per salvare il pianeta dobbiamo farla finita con il capitalismo*. Garzanti, Varese.
Kendra, J., & Wachtendorf, T. (2003). *Disasters: Elements of resilience after the World Trade Center disaster: Reconstituting New York City's Emergency Operations Centre*.
Kerschner, C. (2010). Economic de-growth vs. steady-state economy. *Journal of Cleaner Production, 18*, 544–551.
Kimhi, S., & Shamai, M. (2004). Community resilience and the impact of stress: Adult response to Israel's withdrawal from Lebanon. *IEEE Journal of Community Psychology,32*(4), 439–451.
Koch, M. (2013). Welfare after growth: Theoretical discussion and policy implications. *International Journal of Social Quality,3*(1), 4–20.
Korten, D. (2000) *The post corporate world: Life after capitalism*. Berrett-Koehle.
Kostakis, V., Liarokapis, M., Latoufis, K., & Bauwens, M. (2016). The convergence of digital commons with local manufacturing from a degrowth perspective: Two illustrative cases.*Journal of Cleaner Production* (in press).
Krugman, P. (2011). Degrees and dollars. *New York Times*.
LaRouche, L. (1983). *There are no limits to growth*. New Benjamin Franklin House.
Latouche, S. (1979). *Critique de l'impérialisme. Une analyse marxiste non léniniste de l'Impérialisme*. Anthropos, Paris.
Latouche, S. (1993). *In the wake of the affluent society: An exploration of post-development*. Zed Books.
Latouche, S. (1995). *The westernization of the world: Significance, scope and limits of the drive towards global uniformity* (1st ed.). Polity.
Latouche, S. (1998). *L'Autre Afrique: Entre don et marché*. Albin Michel.
Latouche, S. (2004). Degrowth economics. *Le monde diplomatique*. https://www.jussemper.org/Resources/Economic%20Data/Resources/Degrowth%20economics,%20by%20Serge%20Latouche.pdf
Latouche, S. (2004). *Decolonizzare l'immaginario. Il pensiero creativo contro l'economia dell'assurdo*. EMI.
Latouche, S. (2006). *The globe downshifted*. Le monde diplomatique.
Latouche, S. (2007). *La scommessa della decrescita*. Feltrinelli.
Latouche, S. (2009). *Farewell to growth*. Polity Press.
Latouche, S. (2011). *Come si esce dalla società dei consumi. Corsi e percorsi della decrescita*. Bollati Boringhieri.

Latouche, S. (2012). Can the Left Escape Economism? *Capitalism Nature Socialism, 23*(1), 74–78.

Latouche, S. (2013). Prefazione in Bonaiuti, M *La grande transizione: Dal declino alla società della decrescita*. Bollati Boringhieri, Torino.

Latouche, S. (2016). *La decrescita prima della decrescita. Precursori e compagni di strada*. Bollati Boringhieri.

Lavanco, G., & Novara, C. (2012). *Elementi di psicologia di comunità. Progettare, attuare e partecipare il cambiamento sociale*. McGraw-Hill Education.

Lewin, K., & Cartwright, D. (1951). *Field theory in social science*. Harper, New York.

Lietaer, B. (2005/2006, Rapport). *Des monnaies pour les communautés et les régions biogéographiques: un outil déci- sif pour la redynamisation au XXIème siècle*, in Blanc J., *Exclusion et liens Financiers, Monnaies sociales, Economica*, 76.

Lietaert, M. (2007). *Cohousing e condomini solidali. Guida pratica alle nuove forme di vicinato e vita in comune con allegato documentario "Vivere in cohousing"*. Editrice Aam Terra Nuova.

Lietaer, B., & Kennedy, M. (2008). *Monnaies régionales. De nouvelles voies vers une prospérité durable*. Mayers, Editions Charles Leopold.

Likavcan, L., & Scholz-Wackerleb, M. (2016). Technology appropriation in a de-growing economy. *Journal of Cleaner Production* (in press).

Lovelock, J. E. (1979). *Gaia: A new look at life on earth*. Oxford University Press.

Maggio, G., & Cacciola, G. (2012). When will oil, natural gas, and coal peak? *Fuel, 98*, 111–123.

Maier, M. (1996). Architecting principles for systems of systems. In *Proceeding of the Sixth Annual International Symposium. International Council on Systems Engineering*. Boston.

Malthus, T. R. (1817) [1798]. *An essay on the principle of population*. Murray.

Marra, M. (2015). The city is a bitch: The dematerialization of the Urban. *Journal of Biourbanism, 4*(1–2). https://journalofbiourbanism.org/

Martínez-Alier, J. (2009). Socially sustainable economic de-growth. *Development and Change,6*(40), 1099–1119.

Martínez-Alier, J., Pascual, U., Vivien, F-D., & Zaccai, E. (2010). Sustainable de-growth: Mapping the context, criticisms and future prospects of an emergent paradigm. *Ecological Economics, 69*(9), 1741–1747.

Marx, K 1964, *Economic and philosophic manuscripts of 1844*. International.

Mattauch, L., Hepburn, C., & Stern, N. (2018). *Pigou pushes preferences: Decarbonisation and endogenous values* (CESifo Working Paper no. 7404).

Matthey, A. (2010). Imitation with intention and memory: An experiment. *Journal of Behavioral and Experimental Economics* (formerly The Journal of Socio-Economics), *39*(5), 585–594.

Maturana, H., &Varela, F. (1987). *L'albero della conoscenza*. Garzanti.

Mauss, M. (1999). *Essai sur le don*. Presses Universitaries de France, Parigi.

Mauss, M. (2002). *Saggio sul dono. Forma e motivo dello scambio nelle società arcaiche*. Einaudi.

McDonough, W., & Braungart, M. (2002). *Cradle to cradle: Remaking the way we make things*. North Point Press, New York.

McDonough, W., & Braungart, M. (2003). *Dalla culla alla culla*. Blu Edizioni.

McMillan, W. D., & Chavis, M. D. (1986). Sense of community: A definition and theory. *Journal of Community Psychology, 14*, 6–22.

McMillan, W. D. (1996). Sense of community. *Journal of Community Psychology, 24*, 315–325.

Meadows, D, et al. (1992). *Beyond the limits: Global collapse or a sustainable future*. Earthscan Publications.

Meadows, D, et al. (2006). *I nuovi limiti dello sviluppo. La salute del pianeta nel terzo millennio*. Mondadori.

Mill, J. S. (1911). *Principles of political economy*. Longmans Green.

Miller, W. R., & Rollnick, S. (1991). *Motivational interviewing: Preparing people to change addictive behaviour*. Guilford Press.

Minear, R. H. (2011). *The day the Sun rose in the west: Bikini, the Lucky Dragon, and I*. Hawaii.

Mishan, E. J. (1967). *The costs of economic growth*. Staples Press.

Missemer, A. (2017). Nicholas Georgescu-Roegen and degrowth. *European Journal of the History of Economic Thought,3*(24), 493–506.

Mollison, B., & Holmgren, D. (1978). *Permaculture one: A perennial agriculture for human settlements*. Transworld, Melbourne.

Mooney, C. (2011). The truth about fracking. *Scientific American*.

Morin, E. (1985). *Sociologia della sociologia*. Edizioni Lavoro, Roma.

Morin, E. (1999). *Seven complex lessons in education for the future*. UNESCO Publishing.

Morin, E. (2012). *La via per l'avvenire dell'umanità*. Raffaello Cortina.

Mulligan, M. (2002). *Ecological pioneers: A social history of Australian ecological thought and action*. Cambridge University Press.

Mulligan, M., & Nadarajah, Y. (2008). Working on the sustainability of local communities with a "community-engaged" research methodology. In *IEEE Local Environment: The International Journal of Justice and Sustainability*/abstract content, *13*(2).

Munari, B. (1997). *Arte come mestiere*. Editore Laterza, Roma.

Muraca, B., & Döring, R. (2018). *From (strong) sustainability to degrowth: A philosophical and historical reconstruction*. Routledge.

Muraca, B., & Schmelzer, M. (2017). Sustainable degrowth: Historical roots of the search for alternatives to growth in three regions. In I. Borowy & M. Schmelzer (Eds.), *History of the future of economic growth: Historical roots of current debates on sustainable degrowth* (pp. 174–198), Routledge, London.

Nakamura, T., & Kijima, K. (2009). System of system failures: Meta methodology for IT engineering safety. *Systems Research and Behavioral Science, 26.*

Navarro, V., & Torres Lo'pez, J. (2014). *Un proyecto economico para la gente.* https://web-podemos.s3.amazonaws.com/wordpress/wpcontent/uploads/2014/11/DocumentoEconomicoNavarroTorres.pdf.

Nelissen, N., Van Der Straaten, J., & Klinkers, L. (1997). *Classics in environmental studies. An overview of classic texts in environmental studies*, International Books, Utrecht.

Nienaber, M. (2019). *Germany's Chinese growth engine stalls.* https://www.Reuters.com

Nisbet, R., & Perrin, R. G. (1977). *The social bond.* Knopf, New York.

Norris, F. H., & Stevens, S. P., Pfefferbaum, B., & Wyche, K. F., & Pfefferbaum, R. L. (2008). Community resilience as a metaphor, theory, set of capacities, and strategy for disaster readiness. *IEE American Journal of Community Psychology*, *41,* 127–150.

Odum, E. P. (1969). The strategy of ecosystem development. An understanfing of ecological succession provides a basis for resolving man's conflict with nature. *Science*, *164*, 262–270.

Odum, E. P. (1971). *Fundamentals of ecology* (3rd ed.), 1–574, W.B. Saunders Co, Philadelphia.

Palahniuk, C. (1996). *Fight club.* Mondadori.

Pallante, M. (2007). *La decrescita felice. La qualità della vita non dipende dal pil.* Editori Riuniti.

Pauli, G. (2014). *Blue economy: 10 Anni, 100 Innovazioni, 100 Milioni di posti di lavoro.* Edizioni Ambiente.

Paquot, T. (2007). *Elogio del lusso. Ovvero l'utilità dell'inutile.* Castelvecchi.

Pigou, A. C. (1920). *The economics of welfare.* Macmillan.

Piketty, T. (2013). *Capital in the twenty-first century.* Harvard University Press.

Pimentel, D., Wilson, C., McCullum-Gomez, C., & Huang, R. (1997). Economic and environmental benefits of biodiversity. *BioScience, 47*(11), 45–55.

Perrow, C. (1999). *Normal accidents living with high-risk technologies.* Paperbacks.

Polanyi, K. (1994). *La grande trasformazione. Le origini economiche e politiche della nostra epoca.* Einaudi.

Polanyi, K. (2001). *The great transformation: The political and economic origins of our time* (2nd ed.). Foreword by Joseph E. Stiglitz.

Prati, G., & Pietrantoni, L. (2009). *Psicologia di comunita. Psychofenia, XII*(20).
Prochaska, J. O., & DiClemente, C. C. (1982). Transtheoretical therapy: Toward a more integrative model of change. *Psychotherapy: Theory, Research & Practice, 19*(3), 276–288.
Prochaska, J. O., & DiClemente, C. C. (1992). *The transtheoretical approach.* Oxford University Press.
Rappaport, J. (1977). *Community psychology: Values, research and action.* Rinehart & Winston.
Redclift, M. (2002). *Sustainable development: Exploring the contradictions.* Routledge.
Reid, L. (1962). *The sociology of nature.* Penguin Books.
Reserve Bank of Australia. (2019). Annual Report 2019. https://www.rba.gov.au
Ronan, K. R., & Johnston, D. M. (2005). *Promoting community resilience in disasters: The role for schools, youth, and families.* Springer.
Sachs, W. (1999). *Planet dialectics-explorations in environment and development.* Fernwood.
Sahlins, M. (1972). *Stone age economics.* Routledge.
Sahlins, M. (1980). *L'economia dell'età della pietra: scarsità e abbondanza nelle società primitive.* Bompiani.
Sarason, S. B. (1974). *The psychological sense of community.* Jossey Bass.
Serranò, A., & Fasulo, F. (2011). *L'intervista come conversazione. Preparazione, conduzione e analisi del colloquio di ricerca.* Carocci.
Sen, A. (2011). *The idea of justice.* Harvard University Press.
Sharma, R. (2019). *Small is the new big thing economic forecasters got the 2010s wrong. What will the next decade bring?* https://Nytimes.com
Shiva, V 2000. *Stolen harvest: The hijacking of the global food supply.* South End Press.
Siebenhüner, B. (2000). Homo sustinens—Towards a new conception of humans for the science of sustainability. *Ecological Economics,32,* 15–25.
Smith, A. (1776). *An inquiry into the nature and causes of the wealth of nations.* Strahan & Cadell.
Smith, A. (1759). *The theory of moral sentiments.* Liberty Found.
Solow, R. (1974). Intergenerational equity and exhaustible resources. *Review of Economic Studies, Symposium on the Economics of Exhaustible Resource, 41,* 29–45.
Sonn, C. C., & Fisher, A. T. (2002). Sense of community: Community resilient responses to oppression and change. *IEE Journal of Community Psychology, 26*(5), 457–472.
Sovacool, B. K. (2014). What are we doing here? Analyzing fifteen years of energy scholarship and proposing a social science research agenda. *Energy Research & Social Science,1,* 1–29.

Stanzione, M., von Glasersfeld, E., & Tagliagambe, S. (1994). *Quattro dialoghi su scienza e filosofia*. Odredek, Roma.

Steffen, A. (2009). Planetary boundaries and the failure of environmentalism. http://www.worldchanging.com/archives/010551.html

Stern, N. (2006). *The economics of climate change: The stern review*. Cambridge University Press.

Stiglitz, J. E. (2010). *Risk and global economic architecture: Why full financial integration may be undesirable*. Nber Working Paper Series.

Stiglitz, J. E. et al. (2010). Stiglitz report. Reforming the international monetary and financial systems in the wake of the global crisis. *International Monetary and Financial System* appointed by the President of the United Nations. General Assembly, The New Press.

Swaney, J. A., & Evers, M. A. (1989). The social cost concepts of K. William Kapp and Karl Polanyi. *Journal of Economic Issues,33*(1), 7–33.

Tobin, G. A., & Whiteford L. M. (2002). Economic ramifications of disaster: Experiences of displaced persons on the slopes of Mount Tungurahua, Ecuador. *Papers of the Applied Geography Conferences, 25*, 316–324.

Trainer, T. (2012). De-growth: Do you realise what it means? *Futures, 44*(6), 590–599.

Traoré, A. (2002). *Le viol de l'imaginaire*. Actes Sud/Fayard.

Turner, G. M. (2008). A comparison of the limits to growth with 30 years of reality. *Global Environmental Change,18*(3), 397–411.

United Nations. (1973). *Report of the conference on the human environment* Stockholm, 5–16 June 1972, New York.

United Nations. (1992, June 3–14). Agenda 21. *United Nations Conference on Environment and Development*. Rio de Janeiro.

Valentinov, V., & Chatalova, L. (2013). Transaction costs, social costs and open systems: Some common threads. *System Research and Behavioral Sciences,31*(2), 316–326.

van den Bergh, J. C. J. M. (2011). Environment versus growth—A criticism of "degrowth" and a plea for "a-growth". *Ecological Economics, 70*(5), 881–890, Science Direct, Elsevier.

Van Ypersele, J. P. (2015). *Une vie au cœur des turbulences climatiques*. De Boeck supérieur, Bruxelles.

Vatn, A. (1998). Input versus emission taxes: Environmental taxes in a mass balance and transaction costs perspective. *Land Economics,74*(4), 514–525.

Vatn, A. (2005). Rationality, institutions and environmental policy. *Ecological Economics,55*(2), 203–217.

Victor, P. A. (2014). The Kenneth E. Boulding memorial award 2014: Ecological economics: A personal journey. *Ecological Economics, 109*, 93–100, Science Direct, Elsevier.

Walker, B., & Salt, D. (2006). *Resilience thinking: Sustaining ecosystems and people in a changing world* (1st ed.). Island Press.
Waller, V., Farquharson, K., & Dempsey, D. (2015). *Qualitative social research: Contemporary methods for the digital age*. Sage.
Warner, B., & Shapiro, J. (2013). Fractured, fragmented federalism: A study in fracking regulatory policy. *Publius the Journal of Federalism,43*(3), 474–496.
Weart, S. R. (2003). *The discovery of global warming*. Harvard University Press.
Weiss, M., & Cattaneo, C. (2017). Degrowth – Taking stock and reviewing an emerging academic paradigm. *Ecological Economics, 137*, 220–230.
Welch, C. (2018). Why cape town is running out of water, and who's next. https://www.nationalgeographic.com/
White, L. Jr. (1967). The historical roots of our ecological crisis. *Science, 155,* 1203–1207.
World Commission on environment and development. (1987). *Our common future*. Oxford University Press, Oxford.
Yin, R. K. (2003). *Applications of case study research*. Sage.
Yin, R. K. (2009). *Case study research: Designs and methods*. Sage.
Zimmerman, M. A. (2004). *Empowerment theory & adolescent resilience*. European Association for Research on Adolescence.

Index

M. Carcea, *The Road Towards DemocraCity*,
https://doi.org/10.1007/978-981-16-3505-2

D

I

J

T

The manufacturer's authorised representative in the EU is Springer Nature Customer Service Centre GmbH, Europaplatz 3, 69115 Heidelberg, Germany. If you have any concerns regarding our products, please contact ProductSafety@springernature.com

Printed and bound by CPI Group (UK) Ltd, Croydon, CR0 4YY
30/09/2026
02225326-0004